CHASING ICEBERGS

How Frozen Freshwater
Can Save the Planet

MATTHEW H. BIRKHOLD

PEGASUS BOOKS
NEW YORK LONDON

CHASING ICEBERGS

Pegasus Books, Ltd.
148 West 37th Street, 13th Floor
New York, NY 10018

Copyright © 2023 by Matthew H. Birkhold

First Pegasus Books edition February 2023

Interior design by Maria Fernandez

Library of Congress Cataloging-in-Publication Data is available.

ISBN: 978-1-63936-343-8

10 9 8 7 6 5 4 3 2 1

Printed in the United States of America
Distributed by Simon & Schuster
www.pegasusbooks.com

For Jordan

Contents

Prologue

I t is difficult to know what Prince Mohamed Al-Faisal thought when he walked across the flat campus of Iowa State University in early October 1977. The Saudi Arabian magnate was convinced by his friend, the nuclear engineer Professor Abdo Husseiny, that Ames was an ideal place for the prince's revolutionary conference. The sugar maples on campus were glowing bright red and the swamp oaks were turning orange. Nearby, farmers were driving combines to harvest the remaining corn for the season. Just one day earlier, over three thousand visitors had gathered in the campus Dairy Farm Pavilion to cheer on participants in the milk maid contest as they competed to obtain the most milk from their assigned cows. A beauty queen, an animal science professor, and the campus farm herd manager crowned the winner, who would reign on campus for the year. The prince was squarely in the American Midwest, impossibly far from any ocean, but he was here to talk about icebergs.

Al-Faisal, colloquially known as the "Water Prince," had organized a conference with a grandiose title: "The First International Conference and Workshop on Iceberg Utilization for Fresh Water Production, Weather Modification and Other Applications." After working for fifteen years at the Saline Water Conversion Corporation in Saudi Arabia, a company owned by his government, the broad-shouldered forty-year-old was now president of a private business, Iceberg Transport International, Ltd. He was determined to gather the best minds to figure out how to tap into this frozen freshwater resource. Participants came from every continent, except Antarctica, including glaciologists from Australia, engineers from France, researchers from Libya, and a venture capitalist from Monaco. Even the luminary scientist, Joanne Simpson, the first woman in the United States to earn a PhD in meteorology, attended to help tackle the challenge.

On campus, the conference caused a stir. Clad in a dark suit and a crisp white shirt, the Water Prince cut a fine figure walking around the neoclassical façade of the Memorial Union. He graciously posed for pictures with perfectly coiffed hair and a sparkling smile. The real star of the event, though, was an Alaskan iceberg. One month earlier, a diver plunged into the chilly waters off the coast of Anchorage to wrap an iceberg in polyurethane with the help of a plastics expert. They selected a six-foot-long, 4,785-pound berg, about the size of a white rhino. Once the diver safely insulated the berg, he wrapped a heavy sea fiber cable net around the mass and attached it to a rope dangling from a US Arctic Naval Research Laboratory helicopter. Plucked from the water, the iceberg went straight to the Anchorage airport, where it was packed in dry ice and Styrofoam and put on a plane to Minnesota. From Minneapolis, it was loaded onto a freezer truck and driven more than three hours south to Ames, passing over a landscape long ago flattened by glaciers, which left behind the rich soils tilled by Iowan farmers today. On campus, the iceberg was stored in a walk-in freezer in the Memorial Union.

The process certainly did not make iceberg harvesting seem like an easy endeavor. Dan Zaffarano, then vice president for research at Iowa State University, nonetheless thought the result was worth the effort. He explained at the time, "We felt it was needed for those persons attending the conference who have never seen an iceberg before." Prince Mohamed, in fact, had himself never seen an iceberg. And he had virtually no experience in the Arctic or Antarctica. He had something better: an imagination. At the conference, alongside the scholarly lectures, the prince presented a paper. With the help of a technical adviser, he pitched an innovative method for transporting icebergs. Attach paddle wheels, operated by individual power plants, to a berg. That way, the icy behemoth would become a self-propelled, self-contained unit that could navigate to any destination.

Privately, some conference attendees laughed. The Water Prince's idea was absurdly unfeasible both in terms of engineering and cost. Others dismissed the entire premise of the conference as nothing more than fantasy. The former director of the US Army Cold Region Research and Engineering Laboratory, Dr. Henri Bader, offered a sobering warning to those in the audience: "You engineers should be horrified. You are being asked to develop a technology, which in essential respects lies several orders of magnitude beyond anything within your experience." A number of attendees had come only in hopes of impressing the prince to secure funding to conduct their own polar research. Iowa governor Robert Ray attended the formal dinner that crowned the conference, too. He sidled up to the prince at the head table, sipping a drink chilled with ice cubes chipped from the Alaskan berg, which sparkled as the evening's centerpiece. Dr. Olav Orheim, who would go on to become the director of the prestigious Norwegian Polar Institute, remembers "it was a spectacle, just for show. We did not know what we were doing."

The optimistic prince had a different take. On the penultimate day of the conference, a yellow forklift drove through the Memorial Union toward the berg. It scooped up the mass, maneuvered it through the columned hallways, and dumped the ice outside. The prince walked out of the building carrying a piece of the iceberg in his bare hands. He lifted it so high over his head, his jacket sleeves pooled to his elbows and his shoulders lifted toward his ears. A giddy smile spread across his face. Al-Faisal declared: "We can definitely say the iceberg project is feasible. The only question is when we can begin. Within three to five years, we think we can have a towed berg in situ."

Onlookers cheered. Over the next few days, college students passed the enormous chunk of ice on their way to class and elementary pupils visited campus to admire the ice and touch its frozen surface. One local woman even came armed with an ice pick, plastic bags, and a bucket. As her husband snapped photographs, she chopped into the berg, splintering small pieces of ice that she stuffed into the bags. For later use at a cocktail party, she explained. In the end, the iceberg slowly melted, leaving a soggy patch on the lawn on Harvester Plaza.

Participants left Ames with mixed emotions. It was unclear what would come of the lofty discussions. The conference could very well turn out like the Alaskan iceberg. A tremendous effort for a brief pageant. Exciting but ephemeral. Had the prince done enough to convince people to undertake the effort?

Just six months later, on the other side of the world, crowds gathered in Sydney Harbor to welcome an iceberg. Dick Smith, a thirty-four-year-old electronics entrepreneur, had been studying for months the best way to tow an iceberg from Antarctica. Finally, on March 31, 1978, Smith announced that he would beat the Water Prince and an iceberg would be arriving imminently. The next day, the switchboards jammed as reports came in that an iceberg was spotted. People thronged to the steep cliffs of the south headlands, the traditional viewing grounds of the famous Sydney to Hobart Yacht Race, to catch a glimpse of the ice as it snaked into the harbor. The Royal Australian Navy soon inter-vened, calling Smith's company to offer a place for the ice to moor on the misty Saturday morning. The young entrepreneur could not be reached. He was busy on the barge towing the iceberg. Dressed in a stylish suit, Smith waved and smiled for the cameras that rushed to capture the sight. Soon, boaters approached Smith, begging for pieces of ice. But as the mist turned to rain, the iceberg started to melt. With every drop, more and more vanished and the city began to sober up.

Sydney had fallen for an elaborate April Fools' joke. For just over $1,000, Smith hired a barge and bought a large sheet of white plastic. At 3:00 A.M., under the cover of darkness, the schemer and his team tugged the barge out to sea and created the berg. They layered firefighting foam over the plastic and supplemented it with ordinary, commercially available shaving cream. After two hours of squeezing dozens of cans, they had built the berg and Smith instructed his employees to start calling local newspapers and radio stations about the sight moving through the Sydney Heads. Whether the crowd had been conditioned by media reports or was desperate enough for new water sources, they were ready to see the lump of foam as glistening ice.

The farce made clear what many at that first iceberg utilization conference had feared. In the late 1970s, the notion of using icebergs for freshwater was, at worst, nothing more than an embarrassing jape. At best, it was the stuff of science fiction. The endeavor to understand iceberg harvesting soon slipped out of the popular imagination and remained mostly hidden from view. A few dedicated scientists, including heavy hitters at NASA and the likes of Dr. Orheim, however, continued to pursue the effort, this time with a new crop of entrepreneurs and visionaries. Today, they say that utilizing icebergs as a freshwater source should no longer be treated as a joke. It is time to look below the surface, beyond the spectacle and skepticism the idea initially inspires, to see how substantive the hulking promise of iceberg harvesting might truly be and what it will take to get there.

Introduction
A Quixotic Fantasy?

I curse my husband as my foot slips and sprays loose stones down the path behind me. He has convinced me to climb Table Mountain, the iconic sandstone backdrop of Cape Town, South Africa. From a distance, it is obvious why the Portuguese navigator António de Saldanha gave the mountain its name. The enormous plateau, stretching nearly two miles, is strikingly level and drops off into sheer cliffs. At 3,500 feet tall, it towers over the adjacent bay like a colossal tabletop. Up close, though, I'm focused on the dusty finger grips before my eyes. I'm determined not to let my fear of heights spoil this experience. Jordan has bounded ahead, leaving me sweating behind with our guide, Riann. As a distraction, I ask him how to say *iceberg* in Afrikaans, one of the eleven official languages in South Africa. "Ysberg," he croons, rolling his *r* and exaggerating the fricative *g*. The answer shouldn't come as a surprise, since Afrikaans is based on seventeenth-century Dutch. At that time, when Rembrandt dominated the art scene and tulips captivated the financial market, the Dutch ruled the whale trade, supplying most of Europe with oil for lamps and whalebone for corsets and hoopskirts. The legacy of this Golden Age dominance is recorded in our language: in addition to maritime words like *maelstrom*, *skipper*, and *cruise*, the term *iceberg*—from the Dutch *ijsberg*, or "ice mountain"—stems from whalers' exploits in the Arctic. Today, languages throughout the world use a variation of this word, including Albanian, Russian, Spanish, Urdu, and Yiddish. The linguistic drift brought *iceberg* to the sun-kissed coasts of South Africa, too.

It is a bright May day and I can't tell behind Riann's dark shades whether he's rolling his eyes. Icebergs feel a world away. Along winding roads lined by manicured palms, posh glass-and-concrete villas stand along the western edge of Table Mountain. The beach appears just steps away, but I'm stuck on a three-foot ledge looking straight down the face of the mountain. A series of metal rungs and a well-placed chain make the vertical ascent possible for a cautious climber like me. As we hike higher, we twist to the shady side of the national park. The Cape of Good Hope stands due south and the city sprawls to the east, leading to one of the largest slums in the world. Khayelitsha is home to an estimated half million people. A majority live in makeshift homes built from corrugated metal scraps that lack electricity and running water. In the future, life may become even harder for the people who live here.

In 2018 Cape Town officials officially calculated "Day Zero," when municipal waterworks would be shut down. All 4.5 million residents would have to line up for their daily allotment of water, those living in glamorous villas and crude shacks alike. It is all too easy to imagine that some would suffer more under these conditions. Luckily, the crisis was averted thanks to heaven-sent rains, but water shortages remain a constant threat. For that reason, some visionaries have imagined dragging an iceberg to the capital known as "Mother City" to sustain its inhabitants.

From the top of Table Mountain, the scarcity of potable water is undetectable and the profound poverty is invisible. I'm overwhelmed instead by the stunning view of the city cradled between the shrub-covered mountains and the azure ocean. An iceberg floating in Table Bay would sparkle. Each morning, the city would watch the frozen block begin to glow as the sun climbed over Table Mountain. Tourists would dig their feet into the warm sandy beaches and marvel at the ice, too far to reach by swimming yet still an imposing presence on the waves. The view would be marred only by the small water-processing plant nearby, which would be worth it for the steady stream of freshwater flowing from the berg through underwater pipes into city taps. Denizens might grow worried as the supply shrinks—the product of the thirsty city and the beating sun—but soon another iceberg would be towed in to take its place. This one, maybe, a pinnacled mass with jagged peaks and valleys. A nightly stroll along the Victoria & Albert Waterfront could now include the chance to see the setting sun illuminate the lifesaving iceberg from behind. Could such a sight ever be more than a quixotic fantasy? Nearly fifty years after Prince Mohamed's first Iceberg Utilization conference, a team of glaciologists, entrepreneurs, and engineers believes iceberg towing is possible. And they believe Cape Town is the most likely place we'll see it happen.

◆

For centuries, icebergs have captivated our imaginations. Scientists have striven to understand their enigmatic forms, artists and poets have sought to capture their incandescent beauty, and mariners have struggled against their concealed danger. Icebergs are a daily reality for some and a momentary brush with the divine for others. They are also a potential solution to the imminent water crisis.

By 2030 global demand for freshwater will exceed supply by 40 percent; 107 countries will lack a sustainably managed water source and two-thirds of the world's population will face regular water shortages. The recent scare experienced in Cape Town will be the rule and not the exception. Even in cities with abundant water—like Flint, Michigan, and Jackson, Mississippi—sources have already been so badly contaminated that residents must look elsewhere. Already, 1.42 billion people live in areas of high or extremely high water vulnerability. Happily, there is no absolute shortage of freshwater on Earth—a majority of the planet's supply simply remains untapped, hidden from view in some of our planet's least hospitable climes.

More than two-thirds of global freshwater is locked away in ice caps and glaciers. Stuck at the poles in gigantic fortresses of ice, this water remains inaccessible to most. Each year, though, these glaciers produce miraculous parcels of frozen freshwater and send them into the salty oceans as icebergs. Calved from ancient glaciers formed from fallen snow compressed over centuries, icebergs contain some of the purest freshwater on Earth, with almost no minerality or pollutants. For enough water, we only need to harvest these masses before they melt. By the time an average iceberg reaches the eastern coast of Canada, it is the size of a fifteen-story building and contains around two hundred thousand tons of water—enough to meet the daily water needs for one hundred thousand people for two months. More than ten thousand of these icebergs are calved each year. In the Southern Hemisphere the icebergs are even bigger, some weighing billions of tons. In 2000 an iceberg the size of Jamaica calved from the Ross Ice Shelf of Antarctica. A comparatively small 125-million-ton berg, for example, could supply 20 percent of Cape Town's water needs for a year. Harvesting icebergs is thought to be cheaper than desalinating ocean water, and the blocks of ice could be brought to locations where it would be impossible to build a processing plant. People are also far likelier to drink icebergs than reclaimed wastewater. It is time we pay more attention to these lurking giants.

Some entrepreneurs have already begun capturing and collecting icebergs to create small artisanal batches of premium water for affluent consumers. Others are dreaming bigger, imagining water flowing through the Arabian Desert, fed by bergs relocated to the Persian Gulf. They see emerald gardens and fragrant orange groves springing from the red sands. Humanitarians envision mobile iceberg units that can travel anywhere in the world to rescue people in need of water. Still others think we should not do anything with icebergs at all. It is uncertain if the visions can coexist and unclear which should win out.

As climate change intensifies, more people suffer from water insecurity. Urban populations continue to soar, water sources are increasingly contaminated, and water availability is harder to predict. With every degree the planet warms, rainfall becomes more variable. One ill-timed drought could become a monumental tragedy. But climate change has also created new opportunities. As the world warms, the ice caps will produce more and more ocean-bound icebergs. Thanks to advancements in technologies, these once faraway freshwater marvels are capturable by more than just Arctic inhabitants. With increased global attention on this resource and increasing demand for it, a "Cold Rush" for icebergs could soon erupt.

◆

My quest to learn about icebergs began in L'Anse aux Meadows, on the northernmost tip of Newfoundland, Canada. I had traveled with Jordan to the windswept grassland to visit the Viking site established in 1021 C.E.—believed to be the first European settlement in North America outside of Greenland. It was a cold August day dogged by incessant drizzle, and a thick gray mist clung to the coast. It seemed like a longship could slide ashore any moment. After exploring the sod huts and examining the archaeological evidence, our feet were soaked, and I promised Jordan a drink. We found the only restaurant in town in a small wooden house that doubled as an art gallery. There, we ordered the Vinland martini, named after the Viking designation for this remote corner of the world. To our great delight, the cocktails arrived at our table gently fizzing. The jagged pieces of ice swirling around the vodka and partridge berries had been chipped off an iceberg and now crackled as the millennia-old air trapped inside escaped. Inadvertently, the martinis redirected my attention from Vikings to icebergs. As a lawyer and cultural historian who specializes in property, I was fascinated by the provenance of the ice. According to the bartender, he had hopped onto his Jet Ski earlier that day, scooped up the berg from the Labrador Sea, and brought it back to the restaurant. It was that easy. But was it legal? I began to research the ownership of ice and learned about schemes to recreate the bartender's act on a much bigger scale. Then I was hooked.

Since leaving the waterlogged peat bogs of northern Newfoundland, I have immersed myself in the world of iceberg harvesting. Thanks to wide-ranging conversations around the world with scientists, entrepreneurs, diplomats, activists, and everyday people from teachers to fishers, I have become an unwitting expert on the legal, political, and cultural aspects of these icy wonders. Now I hope to be like Virgil—the fictional version of the ancient Roman poet who guides Dante through the underworld in the *Inferno*. Throughout the fourteenth-century epic poem, Virgil offers spiritual instruction and protection until the heroes reach the center of hell—an enormous block of ice in which the vilest sinners stand frozen for eternity. Like Virgil, I mean take you on a journey to fascinating people and places, and I do have an agenda of sorts. However, I do not intend to personify all-knowing wisdom—even as I share what I have learned—because many of the questions I will pose along the way do not have straightforward answers. And you will soon see that I am not leading you through hell, but perhaps to a paradise of sorts, depending on how we answer those questions.

Is there a world in which icebergs can both be towed to specific locations suffering from water shortages, and harvested and sold like fine wine to wealthy customers? What are the environmental risks of dragging these behemoths through thousands of miles of oceanic habitats, and who gets to decide whether those risks are worth taking? Ought we keep as far away as possible and let icebergs follow their natural course? What makes this tug-of-war a legal and even philosophical matter rather than just a physical one is that unlike gold deposits, this treasure moves. And once icebergs hit international waters, they are fair game for anyone to use.

The stakes are high. The first practices that develop around iceberg harvesting may influence the law itself, since international law takes custom into consideration. If enough private corporations treat icebergs a certain way, it may become the rule. From an international law perspective, there is currently no right or wrong way to handle icebergs; multiple parties can stake conflicting claims. Only our own deeply held cultural beliefs offer guidance. Consequently, how we view icebergs—whether as natural wonders, dangers to avoid, quotidian resources, or mysterious bodies to conquer—affects who we think ought to use them and how.

We are finally standing at the threshold the Water Prince imagined he would soon cross in the 1970s. Advances in technology have made iceberg towing more feasible than ever, and the planet is more desperate for freshwater, making the once-absurd idea seem more and more appealing. We also have a better understanding of our planet, our role in climate change, and the peril we collectively face. In the pages that follow, I will take you across the globe to meet the people invested in the future of icebergs. They may be thousands of miles apart, but like ripples on a pond, their actions will sooner or later influence each other and may reverberate for ages. I will also explore the cultural, environmental, and conceptual insights that underlay our ideas about icebergs. I will explain the questions we need to ask, the perspectives we must consider, and the concerns we must balance for this utopian quest to become a reality, further illuminating how these issues intersect with global politics, Indigenous rights, popular culture, climate change, and the law. The next moves we make will not be clearly right or wrong, but they will be consequential. My goal is to arm you with the knowledge to judge for yourself what future we should pursue and whether we should be excited or terrified as we head into it.

To make Al-Faisal's utopian vision a reality, we must clear four major hurdles. First, we must learn to safely approach icebergs and literally harness their value. We will then have to find a way to make this arduous work economically viable and discover how to thwart greedy interests from exclusively controlling the resource. In addition to the technological and commercial considerations, there is hope to save the world with icebergs only if we can ensure that we will not destroy the planet in the process. Finally, we must create an international framework that forestalls conflicts over this resource and ensures it is equitably distributed. The future of icebergs depends not just on the technical and environmental feasibility of harvesting the resource but also on how we balance consumerism with sustainability, local interests versus global needs, and our tolerance for risk in search of reward. It won't be easy.

◆

The Cape of Good Hope got its current name from the fifteenth-century Portuguese mariners who sought to conquer the world. The moniker referred to the explorers' expectant outlook after they identified a direct sea route from Europe around the southern tip of Africa to India and the East. They could finally dominate the trade in cinnamon, cloves, and pepper. The globe, from their perspective, was newly connected and the future was bright.

Standing on the rocky promontory, I feel the same hope. The setting is intoxicating. Flowering sugarbushes, delicate heathers, and hairy grasses spring from the parched sandstone to form a shrubby blanket that softens the landscape. Some plants smell like fine spices; others are richly musky. In the evening, yellow star-shaped flowers growing up scaly tubes emit a sweet honey smell. The Cape is home to over one hundred endemic plant species, meaning they cannot be found anywhere else in the world. Ostriches, cobras, and baboons call the Cape Peninsula home, too. As I hike through the fynbos, I am only worried about the sharp-toothed monkeys, which are known for attacking tourists in search of food. But at the edge of the water, my mind again goes blank. My eyes strain to look toward Antarctica, some 2,500 miles away, and all I see is bright blue ocean. It is an audacious act of hope to imagine that we could drag an iceberg here.

There is a dark side to this place, too. Before it was named the Cape of Good Hope, the craggy bluff was called the Cape of Storms. The cold Benguela Current from Antarctica and the warm Agulhas Current from the Indian Ocean converge at this site to create notoriously rough waters that have sunk countless ships. The most famous calamity comes from a man-of-war that purportedly survived: the *Flying Dutchman*. According to legend, an arrogant captain sought to round the Cape during a storm and swore he would succeed if it took until Judgment Day. The devil heard his vow and the *Flying Dutchman* was subsequently condemned to sail the seas forever. Throughout the eighteenth and nineteenth centuries, mariners claimed to have seen the forsaken ship, reporting that it glowed an ungodly red. Now, sightings of the ghost ship portend imminent disaster.

The true history of the site is even gloomier. European settlers stole the land that makes up the Cape Peninsula from the Khoikhoi and San peoples, valuing it as a convenient halfway station on the sea journey from Europe to Asia. Those same settlers then enslaved people from Angola, Guinea, Mozambique, Madagascar, Indonesia, Bengal, and beyond to farm the land, eventually forming the multicultural community of Cape Town. The slaves, who were beaten and whipped and starved into subordination, undoubtedly did not experience the same optimism felt by the Portuguese and Dutch traders when they realized this parcel of earth connected far-flung places.

Situated at the southwesternmost tip of Africa, the Cape of Good Hope has a dichotomous nature: it is at once a secluded idyll and a busy global crossroads where ocean currents and peoples collide. Since the fifteenth century, this duality has created strange conditions—some good, some bad, depending on one's perspective. Despite the historic recombinations occurring in these waters, dragging an iceberg to the blossoming sugarbushes and frolicking baboons of the Cape would be mixing up the world in a boldly novel way. A glowing iceberg off the coast could be a harbinger of hope or an evil omen like the doomed *Flying Dutchman*. After all, it would not be the first time that the combination of hubris and an iceberg spelled disaster for humankind.

Icebergs themselves are paradoxes. They are sublimely beautiful and terrifying. They are solid but ephemeral. They glisten but are hidden. They are deadly and lifesaving. To properly perceive such objects, we must view them from as many different perspectives as possible. When dealing with them, we must take thoughtful and coordinated strides; otherwise, the looming Cold Rush risks becoming a free-for-all. Like other resource booms, the hunt for icebergs may result in widespread environmental harm, the inequitable distribution of goods, the exploitation of Indigenous peoples, and social and political strife. We must either work together to find a common solution or fight to see who gets to control these ever-valuable resources. Depending on which vision of the polar marvels triumphs, icebergs may be saviors for our quickly warming planet or portents of more environmental conflict. We are standing at a threshold, and the next steps we take will have consequences for generations to come.

1

Battling Bergs

The door buzzes and a voice crackles over the intercom: *identify yourself.* Instinctively, I straighten my tie and comb my fingers through my hair as I nervously pronounce my name. I wasn't expecting security. Just minutes earlier, I stepped off a packed train from New York City and walked past the small shops lining Bank Street in downtown New London, Connecticut. It looks like a typical New England port town. Ferries chug along the coast, the air smells like the ocean, and seagulls squawk overhead. The redbrick building I had approached looks as inconspicuous as a dentist's office. Only the adjacent Fort Trumbull hints that I am someplace extraordinary.

Originally built in 1777 to defend Connecticut from British forces during the American Revolutionary War, the fortification has since been used by several branches of the US military. Union soldiers trained inside its pale granite bastions during the Civil War. During Prohibition, Coast Guard cutters chased rumrunners in Long Island Sound from the New London Base. Later, the Naval Underwater Sound Laboratory developed submarine sonar systems here to fight German U-boats. The fort has since been converted into a public park, and today the Coast Guard occupies a new facility just north. Here, a team is busy fighting an invincible menace that has been plaguing the Atlantic Ocean for far longer than Fort Trumbull has stood on its shores.

The door clicks open and I step inside the International Ice Patrol (IIP). I've come to learn just how treacherous the "mariner's ancient enemy," as one 1960s report put it, really is. Thucydides, the classical Greek author of *The Peloponnesian War* and one of the fathers of Western history, wrote his account of the war between Sparta and Athens because "the events which happened in the past . . . will, at some time or other and in much the same ways, be repeated in the future." Ever since, historians and philosophers like Thomas Hobbes have argued that wise people study the past "to bear themselves prudently in the present and providently towards the future." By this logic, if we're going to plunge into an endeavor as seemingly harebrained as iceberg harvesting, if we're going to battle this enemy, we should first learn from history.

We have read about the RMS *Titanic* and watched cinematic re-creations of ice bending steel, but most of us have become unmoored from the sea and the dangers it contains. We fly instead of cruise across the oceans. Global exploration is conducted with satellites and algorithms, not intrepid captains and fearless crews. We no longer depend on whale blubber for fuel, cosmetics, and soap. Nevertheless, we are still incredibly reliant on the oceans for commerce—more than fifty thousand merchant ships are crisscrossing the waves at any given time—and all indicators suggest that as the planet warms, we will increasingly turn to the oceans for food, energy, and natural resources. I am skeptical that icebergs are still dangerous. I cannot recall any recent iceberg catastrophes and, frankly, icebergs seem more charming than deadly in the twenty-first century.

Michael Hicks, the chief scientist at the International Ice Patrol, greets me. Though he looks like a man who has spent his entire career with the US Coast Guard—cropped hair, broad shoulders, straight posture—out of uniform he looks more like an affable science teacher who knows his job is fascinating. Before stepping down into this calmer role in 2007, Hicks spent years as the commanding officer of the IIP. "How about a tour?" he asks. The trained oceanographer speaks with military precision but with the warmth of a diplomat.

Up a flight of stairs, the hallway is lined with pictures of icebergs and the IIP's emblem. In heraldic terms, it is a tierced per pall reversed shield, meaning it is divided into three parts with a large triangle on the bottom. In the top left panel, a four-engine turboprop surveillance aircraft flies before three vertical red stripes; the US Coast Guard emblem, with its weighty anchors, occupies the top right panel; and grounding the shield, an enormous pinnacled iceberg emerges from cobalt water. There is little ambiguity about the IIP's mission to "monitor the iceberg danger in the North Atlantic Ocean and provide relevant iceberg warning products to the maritime community."

Hicks shows off a library full of books about icebergs. "We're all kind of nerds," he admits, "and obsessed with icebergs." The members of the International Ice Patrol are mostly young Coast Guard petty officers here for a rotation, though many hold master's degrees in oceanography and all are serious about their mission. Almost everyone at the IIP describes the group as "very mature." The International Ice Patrol is actually spread over several locations and by international agreement administered by the US Coast Guard. The Ice Reconnaissance Detachment—the people who fly the plane depicted in the emblem—works out of Newfoundland, Canada. Information is disseminated from the Communications Command in Chesapeake, Virginia. And headquarters are here at the Coast Guard Research and Development Center in New London, Connecticut. These are the great minds keeping the world safe from icebergs.

Marcus Hirschberg sits in a corner office overlooking the Thames River. He's sporting the operational dress uniform of the Coast Guard: a navy blue long-sleeved shirt over matching pants. His name is sewn over the right pocket, otherwise a civilian like me could not pick him out as the commander of the unit. Hirschberg is equally modest himself and praises his brilliant team for the success of the IIP. As they are busy crafting their "product," the Daily Iceberg Limit—a detailed maritime chart that demarcates a precisely zig-zagging line beyond which ships should not sail—Hirschberg advocates for the unit and speaks to the public. He is a great face for the IIP. After studying Spanish literature at an elite liberal arts college, the Manhattan native was working as a software consultant when the 9/11 attacks killed almost three thousand people. He enlisted in the Coast Guard the following August. Since then, Hirschberg has become an intelligence officer intent on making the world a safer place. "I like icebergs," he ruminates, removing his wire-frame glasses, "they're not trying to outsmart you and they can't adapt to your methods like drug smugglers might." But, he stresses, icebergs are lethal.

If icebergs were a weapon of war, they would be one of the best. They are strong enough to sink an aircraft carrier and can travel as far as an intercontinental ballistic missile without making a suborbital spaceflight. Instead, icebergs swim through the oceans for years, moving so erratically their paths are difficult to predict. Colliding with an iceberg in the ocean would be like running into a brick wall—or, depending on its size, like ramming into the Empire State Building. To make matters worse, icebergs are dynamic forces. Ocean water melts the mass from below and the sun beats down overhead, causing icebergs to lose their equilibrium and tip over. As early as the twelfth century, the Danish historian Saxo Grammaticus warned sailors to steer clear of icebergs, lest they be crushed by a rolling mass. More recently, glaciologists have calculated the power of capsizing icebergs. Depending on its dimensions, a berg can release as much energy as a 5.0 magnitude earthquake. Scarier still, icebergs can be almost impossible to detect. Famously, 90 percent of an iceberg lurks beneath the waves. Smaller bergs might only stand a few feet above the ocean's surface, but can still weigh thousands of tons and easily sink a ship. From a distance, icebergs can be difficult to distinguish from fishing boats, particularly in bad weather. Mostly hidden from view and constantly shape-shifting, these long-lived warheads masterly evade detection.

During ice season, which runs from February to July, the rooms of the International Ice Patrol buzz with activity. Petty officers collect information from reconnaissance teams, crunch data on computers, and coordinate with ships at sea. Sometimes they get urgent requests. One June, Hicks answered a call from the annoyed captain of an ocean liner. The International Ice Patrol had advised all ships crossing the Atlantic to stay out of an iceberg-riddled swath of the ocean about the size of Pakistan. The ocean liner was sailing straight toward it. Navigating around the area would have added time to the journey and delayed the passengers' arrival in New York. Each had paid thousands of dollars to sail aboard the ship, which boasts the fastest and most luxurious transatlantic crossing. The captain only needed to cross a small corner of the danger zone established by the IIP, and he was frustrated by their guidance. Couldn't he just zip across?

The captain was motivated by the same combination of greed and hubris that has driven generations of skippers to seek top speeds on the seas: promises to consumers, accountability to shareholders, and shatterproof pride. The threat to their ambitions has also always been the same: icebergs. Jarringly thrust into the present, these ancient blocks of ice, created well before ships were invented and capitalism could be critiqued, have a tragic history of getting in the way.

◆

In 1855 the *Guiding Star* steamed across the Atlantic Ocean en route from Liverpool to Melbourne. Gold had been discovered in Australia in 1851, and now people across the world were flocking to the continent. The *Guiding Star* promised to get them there fast. On this particular trip, a large number of women and children were on board, including Harriet M'Intyre and her three children. After passing the Cape Verde islands, the M'Intyre family scrawled a note and stuffed it in a bottle. They were all in excellent health, they wrote, and the ship was remarkably comfortable. Could the finder of the bottle please forward the message to their friends back home in Scotland? A month later, on March 25, 1855, a slave on his way to St. Antonio Grande found the bottle on the beach. The Scottish addresses were no longer legible, but newspapers around the world soon reported the found letter because it was the last anyone heard from the ill-fated passengers aboard the *Guiding Star*.

Equipped with three lofty masts, each bearing multiple sails, the *Guiding Star* raced across the waves at an impossible speed. The clipper was heading south, far beyond the southernmost latitude other ships would risk, in an effort to shorten the long journey to Australia. Its captain wanted to set a record. Along the way, the *Guiding Star* passed northbound ships that recorded an alarming number of icebergs, including a three-hundred-mile "wall of ice" in the Southern Ocean toward which the *Guiding Star* was cruising. Tragically, the massive iceberg or chain of icebergs was shaped like a horseshoe. The *Guiding Star* sailed into the middle of the crescent, thinking it was avoiding the faraway ice on either side. In reality, the clipper was sailing to a dead end. We don't know how the final moments aboard the *Guiding Star* unfolded. Some speculate the ship crashed into the ice and sunk instantly. Others think it sat, trapped, for days, weeks, or months until everyone aboard starved. By August 1856, the Victorian Government Immigration Report summarized, "The cause of this deplorable catastrophe can only be conjectured; in all probability it has been owing to collision with the ice." When people went missing on the ocean, icebergs were always a likely culprit.

Just one year later, the famed American captain Asa Eldridge, who earlier achieved the record time from Boston to Kolkata, set out from Liverpool for New York. He was captaining the SS *Pacific* with some 186 people aboard. When the *Pacific* never arrived in New York, ships were sent to look for her without luck. Captain Asa and all of the passengers vanished. Like in Australia, worried family members could only speculate about what happened. In the summer of 1861, however, the mystery was solved when a bottle washed up on the Outer Hebrides off the Scottish mainland. The message it contained was eerily lucid. *On board the Pacific from Liverpool to N.Y.—Ship going down. Confusion on board—icebergs around us on every side. I know I cannot escape. I write the cause of our loss that friends may not live in suspense. The finder will please get it published.* The author of that note, William Graham, knew in that moment what many at the time held for true: icebergs meant death.

By the end of the nineteenth century, everyone, it seems, was thinking about the danger of icebergs. The American author Morgan Robertson, who worked as a cabin boy from the age of five, began writing sea stories to pass the time. In 1898, after he had worked his way up to first mate, Robertson published the novella *Futility, or the Wreck of the Titan*, which tells the story of the "unsinkable" ship *Titan*. Its owners are so confident in the ship's indestructibility, they decline to equip it with sufficient lifeboats. Then, on one ill-starred April voyage, the unthinkable happens. The *Titan* hits an iceberg, "the only thing afloat she could not conquer," the story rather unartfully foreshadows on page five. The iceberg rises 100 feet in the air, overshadowing the ship and temporarily silencing everyone on board. "The music in the theater ceased," then "the deafening noise of steel, scraping and crashing over ice," and the screams of women and children. Because the novella was written fourteen years before the RMS *Titanic* would sail the route between Southampton and New York, people later claimed Robertson had been gifted with precognition.

In reality, the danger of icebergs to those traveling the oceans was so well-known, it became something of a trope in popular fiction on both sides of the Atlantic. In 1912 the German author Gerhart Hauptmann published a novel titled *Atlantis* about the fictional ocean liner *Roland*, another indestructible ocean liner without lifeboats. The story's hero, Frederick, marvels at the ship. "The things that European civilisation has accomplished are tremendous," he thinks. But prudent Frederick worries that humankind might be overreaching. "Isn't man's courage utter madness?" In particular, he contemplates the obstacles in the ocean. "Who could hope to avoid one of the many derelicts drifting in the fog almost submerged?" Like the fictional *Titan* before it, the *Roland* was doomed. Like Robertson, Hauptmann became something of a sensation for his novel.

Hauptmann would go on to win the Nobel Prize in Literature in November 1912, just months after the *Titanic* sunk. Not everyone thought he was deserving of the award. On November 16, 1912, a reporter for the *New York Times* claimed, "It would be hard to convince any unprejudiced student of literature that Hauptmann is a great writer. He has done nothing . . . that will last through the ages." Yet Hauptmann had his pulse on the public's interest. That's why he was able to apparently foretell the *Titanic* disaster. The idea of an unsinkable ship striking an iceberg was in the air. It was something people were mentally prepared for. Or, at least, so they thought.

The sinking of the RMS *Titanic* was not an historical anomaly, but the culmination of a century-long trend. Between 1800 and 1912, more than 150 ships sank or suffered severe damage due to icebergs. But the April 1912 disaster was especially monstrous. For more than a year leading up to its maiden voyage, newspapers around the world reported on the *Titanic*, noting its grandiose design, luxurious amenities, and novel safety features. The largest passenger ship in the world boasted a swimming pool, squash court, and multiple libraries, and it was outfitted with state-of-the-art automatic watertight doors. Infamously, the ocean liner carried only twenty lifeboats, which could accommodate some 1,200 people. There were around 2,240 passengers and crew aboard.

On April 14, 1912, at 11:39 P.M., Frederick Fleet, a lookout stationed in a crow's nest nearly 100 feet above the deck of the *Titanic*, rang the alarm bell three times and called the bridge. "Iceberg! Right ahead!" he warned. He scrambled down the mast to report to command. Fleet heard the ship grind past the iceberg and saw chunks of ice crash onto the forecastle deck. Though the seaman initially thought it was nothing more than a "narrow shave," the damage was soon obvious. Below the surface, the collision created several gaps in the hull that allowed the ocean to spill in. Within minutes, more water had flooded the ship than could ever be pumped out. At 12:05 A.M., Captain Edward Smith ordered the passengers mustered. Fleet was busy helping women and children onto lifeboats when he was ordered to navigate one to safety himself. He rowed in total darkness away from the sinking ship until he thought they were clear of the anticipated suction.

Just two hours and forty minutes after the iceberg was spotted, the *Titanic* was gone. More than 1,500 people perished, including some of the wealthiest men and women in the world and hundreds of immigrants seeking a new life in America. It was the deadliest peacetime maritime disaster in history.

The sinking of the *Titanic* was bound to create a sensation. The hype leading up to its maiden voyage, the high-profile passengers, the blunders that led to unnecessary deaths, and the sheer scale of the tragedy ensured that the public would care intensely about the disaster. And the dramatic stories of escape and heroic sacrifice guaranteed it would become a staple of popular culture. The fact that the ship hit an iceberg, however, was not itself shocking. The *New York Times'* headline, TITANIC SINKS FOUR HOURS AFTER HITTING ICEBERG, would not have read as a colossal surprise in 1912, but rather as a logical explanation. Icebergs were a familiar foe.

In the months that followed, the iceberg became an increasingly important part of the story. How did it manage to damage the ship? Were warnings ignored? Why didn't the crew see it? The US Senate launched an official inquiry into the sinking of the *Titanic*—this is how we know what Fleet shouted when he spied the ice. He testified for two days before a special subcommittee at the Waldorf-Astoria Hotel in New York City. Fleet was interrogated about his eyesight, why he was not equipped with binoculars, and whether it would have made a difference anyway. The questions were not about whether the iceberg was unexpected, but rather if it could have been avoided. In retrospect, we know Fleet is not to blame. Weather conditions made seeing difficult. At twenty-two knots, the *Titanic* was traveling far too fast to quickly change course. Better quality rivets should have been used. More lifeboats should have been aboard. An earlier warning was ignored on the bridge. And, perhaps most tragically, the area off the coast

of Newfoundland where the Titanic sank experienced an unseason-
ably high number of icebergs in April 1912.

◆

The International Ice Patrol was born in the wake of the *Titanic*
disaster. Public outcry around the world about the tragedy led to
the first International Conference for the Safety of Life at Sea,
held November 1913 to January 1914, where representatives from
the great maritime nations met to address safety standards at sea.
Chapter III of the resulting convention required that parties "take
all steps to ensure the destruction of derelicts in the northern part
of the Atlantic Ocean" and create "a service for the study and
observation of ice conditions and a service of ice patrol." According
to the convention, the United States would manage the service, but
the parties would help run and fund it. These included Austria-
Hungary, Germany, Norway, Russia, England, and Canada,
among others.

It is difficult to overstate the significance of this accomplishment
from the perspective of international peace and diplomacy. Europe
was a veritable powder keg at the time. Arms races, expansionist
policies, and territorial conflicts were exacerbated by broken communications and widespread distrust throughout the continent.
Just six months after the International Conference for the Safety of
Life at Sea met in London to discuss lifeboats and icebergs, Arch-
duke Franz Ferdinand was assassinated. Still, the International Ice
Patrol was born. The participating nations were each invested in
keeping their passengers and cargo safe from the threat of icebergs,
so they invested time and money into combating the problem, even if
they could not agree on much anything else. The International Ice
Patrol is consequently one of the oldest international organizations.
Its importance can be measured by its operation during the world

wars. In 1915 and 1916, as soldiers dug trenches in France, the IIP continued to sail in the Atlantic. Similarly, the IIP persisted with its mission during World War II until the United States officially entered. It is perhaps unsurprising that in 1943, when operations were suspended, a merchant vessel sank after colliding with an iceberg in an area normally watched by the International Ice Patrol.

The idea of spectacular collision with an iceberg seems novel today because we have forgotten how common the incident once was. Our amnesia is a byproduct of the International Ice Patrol's tremendous success. They keep us so safe, we do not think much about icebergs as fatal maritime derelicts beyond the *Titanic* disaster, which consequently feels exceptional.

For more than a century, the International Ice Patrol has protected the world from the danger posed by icebergs. In the first decades of its operation, the IIP's methods were charmingly straightforward. Aboard a two-hundred-foot steam-powered cutter, members of the IIP searched the North Atlantic for icebergs. Once they located the southernmost berg, they would sail alongside until it melted. While watching the ice slowly disappear, the crew would notify other ships in the Atlantic of the iceberg's location and recommend they avoid the area. Once it was gone, they searched for the next southernmost berg and repeated the process. This went on for the entirety of the ice season. It would be much better, Commander Edward H. "Iceberg" Smith, thought, if they could eliminate the threat altogether.

◆

On a calm May day in 1923, a team of International Ice Patrol members spied a prime subject for their dangerous experiment. The pinnacled berg stretched 170 feet into the air, almost as tall as the leaning Tower of Pisa, and appeared sturdy. The sun and ocean had not yet deteriorated the ice, which was deemed to have a lot of life left. A few men jumped into a lifeboat with their equipment and rowed toward the berg. They had carefully lashed two mines together. The heavy steel orbs each contained hundreds of pounds of TNT, ready to detonate when one of the fuses made contact with a solid object. The danger, though, stemmed not from the mines, but rather the ice. Even in the tranquil weather, the berg was considered too erratic to closely approach. It could flip at any minute and crush the crew. The team, accordingly, rowed to the windward side of the ice, where they attached the mines to a floatation device rigged with a small sail and set it in the water about 50 feet away from the target. Suspended underwater, the mines rapidly approached the ice but got caught in the splash back and narrowly missed the mark. A more certain approach was needed. The crew next strung a long line to the float. They would bring the mines to the berg by towing the line across the face. But the waves proved too uneven to control the float from such a distance. Frustrated, the men picked up the mines. For the third attempt, they backed in close to the ice, lowered the mines, and fired. Nothing. The batteries malfunctioned. The crew rowed back to the ship and regrouped. As the hours passed, fog rolled in and a new plan developed. This time, the men would risk getting closer. They hopped back in the rowboat and found an icy ledge projecting from the underside of the berg. The haze was so thick now, the peak of the berg was not visible. Carefully, the men placed the mines over the ledge, pulled away, and fired. An explosion recoiled off the berg. They couldn't see much, but heard as ice came crashing down. Twenty minutes passed and the iceberg could still be heard calving. Hours later, when the fog finally cleared, the team saw three pieces had broken off the berg, which was still an imposing presence on the waves.

The next morning, the team reapproached the mass, where a new sheer wall, 15 feet tall, appeared. After several attempts, the men anchored a grapnel in a small crevasse at the top of the ice face. New mines were attached to the line and lowered into the water. This ensured they would eventually collide with the ice. Another explosion. A jet of water shot into the air and more ice came crumbling down. Meanwhile, the Atlantic was filled with cargo ships and cruise liners. The *Mauretania*, the *Lapland*, and the RMS *Olympic*—the *Titanic*'s sister White Star ship with more than three thousand people aboard—passed nearby in the fog.

After one day off, the team tried a third time. One man, sporting a shoulder gun, sank another hook into the colossus and another mine was attached to the shot line. This time, the berg split in half where the mine detonated. In the end, the US Coast Guard considered the experiment a success. According to the official report, "It is believed that the life of this berg was certainly shortened by more than one day, possibly two." Four days of work and thousands of pounds of TNT had lessened the danger of this iceberg by less than forty-eight hours.

As war technologies evolved, they were systematically applied to icebergs, the mariner's ancient enemy. In 1959, armed with new bombs from World War II, the International Ice Patrol decided to drop clusters on icebergs. The crew selected for the mission practiced for days, dropping dummy bombs onto icebergs from 1,000 feet. When the big day finally came, most of the bombs missed the moving targets, and those that hit mostly bounced off the ice and into the ocean. When one bomb finally successfully struck, a huge explosion resulted. Water sprayed 500 feet into the air as the berg shook. Even so, the IIP measured no material change in size or attitude. The *New York Times* reported that "the icebergs had been more or less impervious to destructive man-made means."

The next year, the International Ice Patrol decided to deploy thermite, a pyrotechnic powder that burns very hot and very fast. A thermal shock, it was thought, would fracture an iceberg along its weakest points and hasten its melting. A team of three crampons-clad men carefully boarded the selected berg. They carried a gasoline-powered ice drill as they gingerly walked across a flat surface, digging their shoes into the ice to avoid falling into the freezing waters. The temperature was so deadly, they did not even bother wearing life vests. If they fell in, the men would almost instantly be killed. The team had to move quickly. Even if their specialized equipment could keep them affixed to the ice, the berg could roll and drag them under the surface. They swiftly drilled, set a 168-pound thermite explosive, and hopped off the berg. Unlike a majority of the underwater mines and aerial bombs, this blast was sure to hit. When the smoke cleared, the team saw the result: a crater no deeper than the handle of an ice axe. The tactic was jettisoned. The same conclusion was reached when it came to the experiments with lampblack, or carbon black. Theoretically, the powdery material collected from oil and gas soot would accelerate the melting of an iceberg. Another crew embarked onto an iceberg and spread one hundred pounds of the toxic substance over a 75 x 100 foot swatch of ice. They checked back twelve hours later, only to learn that most of the black coating had been washed away. The results of these 1960s trials, in the words of Captain Ross P. Bullard, were "negligible" and patently not worth the risk of approaching an iceberg. Even the less forceful undertakings had proven useless. Tracking icebergs by marking them with dye-tipped arrows and radio transponders failed because icebergs constantly turn and melt. It is impossible to try to keep something on or in them.

It is easy to guess what Thucydides would think about the various attempts to conquer icebergs. Brute force is not the way to defeat this enemy. Over the years, the International Ice Patrol tried to annihilate icebergs with torpedoes, naval shells, mines, aerial bombs, thermite, and dynamite. In many instances, these stratagems actually multiplied the danger, since more, smaller, and still-deadly icebergs resulted. The International Ice Patrol soon abandoned the attempts as useless.

◆

I ask Mike Hicks if we've surrendered to icebergs. "Not exactly," he answers, clarifying "we've just learned how to better live with them." Thanks to the International Ice Patrol, this is certainly the case. The safety record speaks for itself. Before the IIP existed, hundreds of ships and thousands of people died because of iceberg collisions. Since 1914, when the first IIP cutter hit the waves, no captain that has followed its guidance has struck a berg. A ship must simply listen to the IIP to guarantee it won't hit an iceberg in the North Atlantic. Just like that, the common literary trope vanished and iceberg collisions became an exotic novelty.

Bafflingly, every few years, some imprudent captain chooses to ignore the IIP's sterling advice. The SS *Hans Hedtoft* has the ignominious distinction of being the last known ship sunk by an iceberg with casualties. On the homeward leg of her maiden voyage in 1959, the ship cut through an area the IIP warned to avoid. The diesel-powered freighter had supposedly been designed to withstand polar ice. She would have no problem sailing the North Atlantic, it was claimed. Without worry, women and children boarded the 2,857-ton ship, as did one of Greenland's representatives to the Danish Parliament, but the trip from Nuuk to Copenhagen proved deadly. Twenty-foot waves tossed the ship into an iceberg. All ninety-five people aboard perished. *Time* magazine reported the accident in its very next issue as the "Little Titanic" and pontificated that "once again . . . the cruel sea . . . made a mockery of the vanity of man." The International Ice Patrol and Coast Guard tried to rescue the vessel, only to confront an ocean that was choked with icebergs. The SS *Hans Hedtoft* never stood a chance.

"Not every accident is so dramatic," Hicks reassures. More commonly, ships that ignore the International Ice Patrol are looking to save money. Rerouting to a longer path to avoid icebergs takes extra time, and even a one-day delay can cost a mind-boggling sum. But if a ship strikes a berg, the time and money the captain attempts to save by cutting through the danger zone quickly evaporate. The smartest plan is to listen to the IIP.

Ironically, the perfect safety record itself creates a problem. Hirschberg explains that "it can be hard to demonstrate there is danger because we have not had an incident." The International Ice Patrol, in other words, is so good at its job, we have forgotten how necessary its work is. In reality, icebergs remain a threat and are becoming increasingly problematic as the planet warms. Arctic glaciers can calve more than ten thousand icebergs each year. Sometimes just a few hundred make it as far south as the 48th parallel north. Other years, thousands of icebergs fill the North Atlantic. In 2017, in just one week, 450 icebergs floated off the coast of Newfoundland, up from thirty-seven the prior week. Experts described the icebergs as an invading fleet. It was masterfully evaded, of course, by the IIP.

The record also creates some internal pressure. "I can imagine not wanting to be the person to lose the streak," Hicks admits. "Why not lie, then?" I ask. "If you want to keep the record and keep everyone safe, demarcate an extra-large 'do-not-sail' area. Forget Pakistan, make it as big as China." Hicks patiently explains the tension. To ensure that the IIP's daily ice limit is taken seriously, the patrol cannot be too conservative. Otherwise, mariners would stop believing the product, and they would think that they could perhaps travel a little bit into it, or cut some corners. So the smart mariners I see in the halls of the New London facility have to make the map as close to reality as possible, which is exceedingly difficult since icebergs move.

◆

Luckily, the International Ice Patrol excels at the task at mapping icebergs. Technically, an iceberg is a frozen mass of freshwater that extends at least five meters above the water line and covers an area of at least five hundred square meters. Smaller pieces of ice are known as "bergy bits" and "growlers." Both can originate from glaciers or come from icebergs as they break apart. Bergy bits sound cute, but are deadly. They reach between one and five meters above the water and are typically one hundred to three hundred square meters, or the size of an average single-family home in the suburbs. Growlers reach less than one meter above sea level and are less than two meters across. Ramming a growler in the ocean would be like hitting a pickup truck on the highway.

On average, ten to fifteen thousand new icebergs are born annually. Some years, there might be up to forty thousand icebergs. In the Northern Hemisphere, most icebergs come from Greenland; some originate from glaciers on Arctic islands in Canada, the Norwegian archipelago Svalbard, and Novaya Zemlya in Russia. Large icebergs might travel more than five thousand kilometers from their place of origin, and some have been spotted near the subtropical Azores in the middle of the Atlantic. Fortunately for the International Ice Patrol, only 1–2 percent of icebergs in the Northern Hemisphere travel farther south than forty-eight degrees north, which is about where St. John's stands, and still seven degrees north of the *Titanic*'s final resting spot. The IIP focuses on an area known as "Iceberg Alley," which stretches from Labrador to the southeast coast of Newfoundland. From Greenland, icebergs swirl north along the West Greenland Current deep inside Baffin Bay before they pass through Davis Strait and ride the cold Labrador Current south along the Canadian coast into the broader ocean and busy transatlantic shipping lanes.

These mariners have learned from history and no longer try to blow the things up. Nor do they simply search for the southernmost iceberg. Instead, the International Ice Patrol utilizes satellites and algorithms to create a real-time map of the dangers floating in the North Atlantic. Once icebergs are spotted from space, they are plugged into a map that predicts where they will travel and how quickly. The model also uses ocean buoys and historical data to collect information about the ocean currents, water temperatures, and the height of waves, all of which are important to help predict the rate of iceberg deterioration. The IIP is constantly striving to improve its model, relying on machine learning and a variety of partners for the best data possible, including NASA, the National Resource Council, and the Canadian Ice Service.

The sophisticated technology is still plagued by limitations. Although the computer algorithm can identify possible icebergs in the satellite image, the work must still be checked by a member of the International Ice Patrol, since icebergs and ships can look similar from space. The human eye can better pick out the standard shape of ships compared to irregularly shaped icebergs and use clues like the length of the wake, which can stretch much farther when created by a ship. Hicks tells me that the computers are getting better and better at differentiating the two. Earthly realities create problems, too. The area surveyed by the IIP is covered by clouds for more than 75 percent of ice season. As a result, the satellites often deliver pretty pictures of clouds without giving any hint of the treacherous seas beneath.

The International Ice Patrol consequently relies on reconnaissance flights to make their detailed maps. Aboard an HC-130 Hercules extended-range search and rescue aircraft, IIP members can zoom in on radar or fly below the clouds to visually inspect shimmering objects on the ocean. The missions usually begin by tracing the borders of the current iceberg limit to confirm that no bergs have traveled beyond the limit. If an iceberg is spotted, the crew enters it into a tablet and it is added to the analysis and prediction system to plot the berg's drift and estimate its deterioration. Some years, the IIP examines more than eight thousand potential icebergs. Those that are validated get plugged into the map. Currently, Hicks tells me, the satellites are still missing around 50 percent of the icebergs in the operations area.

The effort is truly collaborative. Funding comes from seventeen different countries, and a variety of organizations and individuals contribute to compiling the best data. A Danish organization provides the iceberg limit near Greenland, and the Canadian Coast Guard shares information. Commercial airlines register icebergs when they fly over Iceberg Alley, as do merchant ships and oil and gas companies that operate in the waters. Even lighthouse keepers report sightings to help make the most accurate map possible.

◆

Back at the International Ice Patrol's headquarters in New London, I ask Hicks whatever happened to the cruise ship that inquired whether it might sail into the danger zone. Ships do not legally have to obey the IIP's guidelines, after all. In the end, after some huffing and puffing, the captain agreed to follow the daily iceberg limit. And, keeping the streak alive, the ship arrived without incident to New York, though maybe a few hours later than its passengers and crew would have liked.

Although landlubbers may have forgotten what a grave threat icebergs pose because of the International Ice Patrol's brilliant work, the icy behemoths still endanger anyone who sails near them. I question Commander Hirschberg about the possibility of iceberg harvesting, and he sits quietly for a moment. "I wish they wouldn't," he says. In his estimation, keeping a safe distance from an iceberg is a nonnegotiable safety measure, and harvesters simply cannot follow this guidance. In my time with the International Ice Patrol, I learn that the lesson is clear and simple: stay away from icebergs. The danger remains the same as when the RMS *Titanic* sunk—we have just better learned how to keep our distance. As a lawyer, I have been trained to assess risk, and as a historian I have been trained to learn from the past. We have battled this enemy in the Atlantic for more than a century, and though the Coasties will not say so, I cannot help but conclude that icebergs have won. Others are much braver than I, though, and more enterprising, too.

2

How to Capture
Frozen Freshwater

For most of the year, Iceberg Alley is gray and cold. The largest city on its shores, St. John's, is known as "Canada's Weather Champion." Among major Canadian cities, the capital of Newfoundland and Labrador is the snowiest, windiest, wettest, and cloudiest, enjoying fewer than 1,500 hours of sunshine each year. Seattle, for comparison, gets 2,200 hours of sun annually. St. John's is so overcast, the difference between it and Seattle is greater than that between Seattle and Tampa. But for a few months, from May to August, the sun breaks through the clouds and warms the freezing waves swirling off the coast.

In this brief window, when Newfoundland relishes nearly half of the sunshine it will absorb for the entire year, icebergs fill the Labrador Sea. The Arctic ice pack undergoes its seasonal melt and Baffin Bay thaws, allowing the frozen mountains to continue their journey toward the Atlantic. Most break off of glaciers on the west coast of Greenland—what glaciologists call "calving." Speakers of a variety of languages, from Afrikaans to Uzbek, use the same word to define the process, as if the icy masses are the living offspring of glaciers. In Albanian, Farsi, and Italian, it is even more explicit:

glaciers "give birth." Across cultures and languages, icebergs are conceptualized like wild cattle or horses roaming the maritime frontier in our rhetorical imagination. It is no wonder, then, that the International Ice Patrol and Canadian Ice Service describe the summertime influx of icebergs as an annual "migration."

This is when iceberg cowboys head to sea. These rough-and-tumble mariners earn their living wrangling icebergs—sometimes to subdue and capture the leviathans, other times to herd the ice in new directions. They are undaunted by warnings issued by the International Ice Patrol. To that end, the brave seafarers make bad role models for academics interested in icebergs and ship captains navigating the North Atlantic. The iceberg cowboys, however, give us a hint of what it will take to harvest icebergs if we are ever going to use them to save the planet. They show us that these frozen beasts are not entirely unapproachable.

◆

One million years ago, when mastodons roamed the Atlantic Coastal Plain in what is today New London, Connecticut, it was snowing in Greenland. We know this because glaciologists have taken ice cores from the center of the country. Using a drill that cuts a circle around a central point, they extract long, skinny tubes of ice. The process can take years, especially in a place like Greenland, where the ice sheet is more than two miles thick and drilling can only be done during the summer. Back in the lab, the glaciologists then analyze the ice to identify the particulates and chemicals captured by the falling snow. Using this information, they are able to reconstruct past climates and determine local temperature, greenhouse gas concentrations, and volcanic and solar activity.

The Greenland ice sheet was likely first formed some three million years ago, but it has grown and shrunk as the planet has warmed and cooled. It currently stretches 1,500 miles long by 680 miles wide and covers 80 percent of Greenland. Because most of the ice sheet is ringed by mountains, glaciers that lie along the coast like Sermeq Kujalleq function as outlets through which ice and water from the ice sheet come gushing out. Like rivers, glaciers are constantly flowing, pushed forward by the weight of their own ice. Sermeq Kujalleq, also known as Jakobshavn Glacier, travels an average of 130 feet in twenty-four hours and calves more than eleven cubic miles of icebergs each year into the Ilulissat Icefjord. Owing to this constant growth and shrinking, the oldest remaining known ice in the Greenlandic ice sheet today dates from the Pleistocene epoch, one million years ago.

Glaciers form when snow builds up and is buried each year by more and more snow. To understand the process, it helps to think of the snowball fights you might have had as a child. At normal atmospheric pressure, ice melts at thirty-two degrees Fahrenheit. Adding pressure, however, can lower the temperature at which this transformation, or solid-to-liquid phase change as we're taught in chemistry class, occurs. Imagine squeezing a fistful of powdery snow. The force of your clasp manages to melt the snow just a bit. When you open your fingers and release the pressure, the snow refreezes into a harder, more solid lump. During a snowball fight, you might repeat this process a few times, adding a bit more fresh snow to the ball in your hand each time, until you create the perfect ammunition. In Greenland during the Pleistocene epoch, as more and more snow piled up, the ice crystals were compressed and recrystallized over centuries, forming rock-hard, crystal-clear glaciers.

To get to this ancient ice, we don't always have to rely on glaciologists and their specialized drills. Instead, we can wait for icebergs to calve from glaciers and come to us; after all, they can travel thousands of miles before melting. For this reason, the famed Scottish geologist Charles Lyell believed icebergs transported boulders around the world. His dear friend Charles Darwin agreed, further suggesting that icebergs were responsible for the dispersion of species. While this theory has since been debunked, the nineteenth-century naturalists were correct that icebergs can act as transport vessels, bringing artifacts and stories from the past with them. In addition to chemicals and air bubbles trapped in snow, plants and animals can get caught on top of glaciers and become part of the historical record. Sometimes the findings are unbelievable. In the nineteenth century, several people reported seeing woolly mammoths frozen inside icebergs. Their shaggy hair and long, curled tusks were purportedly cryogenically preserved inside the ice. Another legend tells of a fully clad Viking encased in an iceberg, still gripping his spear and shield. Every time an iceberg floats by, an ancient piece of the past is carried along with it.

Iceberg ice is not just primeval; it is also pristine. As snowflakes cascade through the atmosphere, they can catch sundry gaseous and particulate matter thanks to their latticework structures. Snow falling today, for instance, might absorb black carbon, mercury, formaldehyde, pesticides, and vehicle exhaust. Some scientists consequently advise children not to eat snow in urban areas. Thousands of years ago, however, the same pollutants were not floating around our atmosphere, so ancient glacial ice is comparatively immaculate. It has far fewer parts per million of impurities than most tap waters contain today. When glaciers calve, this unpolluted ice is packaged in a tidy parcel and launched into the salty sea.

Glaciologists describe icebergs as having a life cycle. They are born and die. Once an Arctic iceberg calves, it will typically live for three to six years, depending on its size and the journey it takes. An iceberg headed to the equator will melt faster than one grounded near the poles. Besides warmer water, icebergs are deteriorated by waves that crash into them and collisions with other icebergs or land. Cycles of thawing and refreezing further create crevasses within the iceberg that can cause a berg to crumble or explode and spawn further icebergs. The final stage of an iceberg's life is especially volatile and difficult to predict. Scientists struggle to precisely foresee the path an iceberg might follow since the shape and size of an iceberg, ocean currents, winds, and waves affect a berg's velocity. In Iceberg Alley, the average drift speed is about 0.5 mph, though some bergs zip along at more than 2 mph. At this point in their lives, Arctic icebergs are prone to capsizing since increased erosion over time results in a loss of stability. In an iceberg's death throes, as some glaciologists describe it, the blocks are also noisy. Icebergs groan and sputter as they break up over open water. Glaciologists talk about "singing" icebergs, too. When bergs scrape the seabed or rub against each other they can vibrate and emit chilling harmonic tremors, like running a finger around the rim of a wineglass. Perhaps it is to be expected that we talk about icebergs like they are alive. Iceberg cowboys try to capture these creatures as they migrate through Iceberg Alley, before they die and all that unspoiled freshwater mixes with the ocean.

◆

Ed Kean is a fifth-generation fisherman known abroad as "Captain Ahab of the Ice." Some Canadians describe Kean as a "Newfie." Depending on whom you ask, the term is either endearing or offensive, meant to characterize people from Newfoundland as quirky or dim-witted. The stereotype is based on the insular life locals lead on the sparsely populated island and the unique accent that has resulted. In this regard, Kean is no exception. To my ears, his speech sounds like the mix of a cheery Irishman and a grunting whaler too busy or too weary to open his mouth any wider. He looks the part, too. His round face is ruddy from years at sea and he has the stout bearing of a man who can stand on a rocking ship. "Boats sink and boats float, so you have to keep on your toes," he snorts.

Despite Kean's easygoing attitude, he is acutely aware of the danger posed by icebergs that float in the waters off the coast of Newfoundland. Many calved from the very same glacier, Sermeq Kujalleq, that produced the frozen torpedo that sank the *Titanic* just 320 nautical miles away. But the risk is worth it to continue living off the sea. Prior to 1992, St. John's, like most other communities in Atlantic Canada, was a fishing town. That summer the Canadian Federal Minister of Fisheries and Oceans declared a moratorium on the devastatingly depleted Atlantic Cod fishery. Seemingly overnight, the economy in Newfoundland and Labrador went belly-up. More than twenty-two thousand fishers and plant workers lost their jobs. The federal government intervened to provide income assistance, retrain Newfies for different employment, and incentivize education. As a result, the economy is better diversified today. Locals now fish for crab and shrimp and drill for oil and gas. Some, like Kean, also set their eyes on the ice giants that swim by every summer, earning paychecks wrangling the bergs for companies interested in capturing some of the purest freshwater on earth.

As the iceberg cowboy tells it, his family used to harvest ice from icebergs to cool fish that they brought up to the plants in Labrador. In the 1970s they had the ice tested at Memorial University of Newfoundland to ensure it was safe. In the process, they learned that it was exceptionally pure. In the 1990s Kean began harvesting ice commercially, and he has been at it ever since.

Icebergs "taste like water should taste," he explains. That's why they are worth wrangling. According to Kean, he and his wife drink several liters of iceberg water a day and his mother "will only drink iceberg water." She even uses it to make her tea and coffee. The Newfoundlander emphasizes his mother's age—eighty-six years old when we speak—though it is unclear if Kean is implying that he pampers his mother like a queen or that the iceberg water is so healthy it has kept the aging woman hale.

In Newfoundland, several local companies rely on iceberg water, including the cosmetics company East Coast Glow, Quidi Vidi Brewery to make beer, and Auk Island Winery to blend berry wines. They do not get the ice themselves. In St. John's, when I tell people I'm interested in iceberg harvesting, they almost instantly reference Ed Kean, who is something of a celebrity here. Several people then tell me a sad story about a man who died a few years back. To get ice for a drink, he hopped in a kayak and paddled out to an iceberg. As the Newfoundlanders recount the story, the mass unexpectedly rolled and trapped the kayaker underneath, where he quickly drowned. All for novelty ice cubes at home. This work is best left to professionals.

The global COVID-19 pandemic spoiled my plan to sail with Kean, so I chat with him about icebergs while we're both standing on solid land. Proudly, the fisherman also directs me to videos online that document his work. By all accounts, it is arduous labor. Some years, Kean sells more than one million liters of the specialty water.

On days he harvests icebergs, which is almost every day in the short migration season, Kean is awake by dawn with a small crew. Using a satellite map, Kean identifies icebergs and motors his large fishing boat toward them. Once he sees the ice, he can select his approach. If the berg is grounded and sufficiently stable, he can carefully maneuver his accompanying barge, outfitted with a crane, near the ice. Then, using a hydraulic clamp attachment, he and his crew scoop up the ice and feed it into a large grinder on the ship. The ice is shredded and piped into storage tanks.

More commonly, the iceberg cowboy cannot get so close to the ice and harvesting looks a lot more like battling a wild animal.

Because the ice can flip and roll unpredictably, Kean must approach cautiously. Plus, icebergs have treacherous underwater projections or "legs" Kean names them, that could damage his ship. Occassionally, his first move is to whip out a rifle. He shoots icebergs in hopes of splintering off a chunk. "Sometimes it works, sometimes it doesn't." Kean tells me that he is moving away from firearms now, though. The crew next leaves the big rusty rig behind and hops into a small motorboat to come near the ice. They circle the iceberg looking for smaller pieces that have broken off. Even these chunks can weigh more than a ton. The next part comes right out of a rodeo. The cowboys use long wooden poles tipped with metal hooks to prod and jostle the ice closer to the boat. Once the iceberg is well positioned, the men fling a net over the ice and wrap it tight. The brute is then dragged back to the vessel, where a crane is lowered to affix a hook to the net. The ice is winched aboard and dropped onto the deck. The ice is rinsed, and now the backbreaking work begins. Using axes, the crew hacks into the ice, chipping it into smaller chunks. These are then shoveled into one-thousand-liter plastic storage containers where they will melt. Kean says he has difficulty maintaining a crew for many seasons since the work is so difficult. Maybe I should be grateful that I don't have to participate in this rodeo.

Since I don't get the opportunity to play cowboy in St. John's, I decide to be a tourist instead. I walk along Jellybean Row and admire the vibrantly colored clapboard homes adorned with gingerbread trim. I climb Signal Hill, overlooking the narrow passage between the harbor and ocean, and scan the horizon to see a glimmer of ice in the bright sun. My eye reaches the place where the sea seems to touch the sky. Mariners call this area the "offing," meaning the ocean beyond anchoring ground. Still no icebergs. I resolve to march down the steep slope and board a ship from St. John's Port Authority to see an iceberg. Seventy-five dollars buys me a two-hour tour. "No icebergs guaranteed," I'm repeatedly warned. I board the bright red trawler and breathe in the fresh Atlantic air.

If harvesting icebergs is like herding cattle, sightseeing for icebergs is itself a bit like being on safari. My tour operator even describes "hunting" for icebergs on the open sea. Our target is dangerous and unpredictable, capable of lashing out at any moment. For that reason, the provincial government of Newfoundland and Labrador has published viewing guidelines. We should keep a distance equal to the length of an iceberg or twice its height, whichever is greater. Anything within this perimeter is considered the "danger zone," in which viewers are exposed to falling ice, large waves, and submerged hazards. As with lions and leopards, it can also be hard to tell where or when our mark might appear. "Who knows what nature is going to give us," my guide explains. Tour companies rely on identified icebergs to drum up business. They tell tourists and other come-from-aways, as outsiders are called here, that they know exactly where to find beautiful icebergs. This advertising can sustain a steady stream of business, since icebergs, unlike large game, cannot easily hide or nimbly slink away once they are spotted. The threat to this boon, I learn, is from harvesters like Ed Kean. While tour operators motor sightseers back to St. John's Port Authority to pick up the next batch, they worry that iceberg cowboys might come along and scoop up the very same iceberg that has resulted in happy customers and good reviewers. When I ask about it, my guide is brusque: "We won't see any of those folks on our tour."

Luckily, I get to see icebergs. The first we encounter rises out of the ocean like a smooth boulder. A cotillion of terns is settled at the peak of the dome's gentle curves, and for a moment, the ice seems passive and gentle. Then a wave crashes into the mass and I am reminded that the frozen surface is not a static island, but an enormous, peripatetic rock-hard reef. As the boat inches closer, I can see the slippery surface continue below the waves, stretching into the dark ocean. I am transported by the alien sight up close. The floating orb is like a puffy white flying saucer cutting through the water. The next iceberg we spy is more spectacular, sparkling like a crystal castle in the bright sun. Two pinnacles, like turret-adored towers, loom fifty feet over our ship. They are connected with a low ice wall that reminds me of a crenellated parapet. I ask one of the guides if she thinks we can sail nearer. "That's up to the captain," she answers, "but he probably won't get closer because this berg isn't grounded." Salt water lashes the ice from every angle, like the ocean is a great artist carving a magnificent sculpture from a giant block. The ice dwarfs our boat even from our safe distance. When I finally look down, I am mesmerized by the aquamarine hue radiating from the submarine foundation. Intellectually, I know that around 90 percent of an iceberg's mass is hidden. But there is so much ice bobbing on the waves, I cannot fathom the true size of this creature.

On our return journey to St. John's we pass a pair of humpback whales gliding through the ocean. Our guide guesses they are about forty feet long and weigh close to forty tons. The enormous mammals shoot vapory jets skyward and flash their shiny black dorsal fins as they undulate on the surface. I reflexively hold my breath as the leviathans swim closer. There is nothing I could do to stop the whales; they could easily crush our ship and fling me into the sea. I move inside the cabin and am thunderstruck as I mentally compare the whales to the one-hundred-thousand-ton icebergs nearby. They might move slower than the cetaceans, but the icebergs are just as powerful and equally sovereign.

Technically, I've seen a domed and a pinnacled iceberg. Glaciologists distinguish between six different types. The categories are helpful because different shaped bergs have greater or lesser

height-to-drift ratios and drift speed as a percentage of wind
speed. Some varieties are also more attractive for harvesting,
depending on whom you ask. Domed icebergs are smooth and
rounded on top. A blocky iceberg has a mostly flat top and steep
vertical sides. Like the name implies, wedge icebergs have one
thick end that tapers to a thin edge that disappears beneath the
water. Pinnacled icebergs can be shaped like pyramids or consist
of several spires that soar above the bulk of the mass. Similarly,
dry-dock icebergs have two points like a giant letter *U*, the base
of which is at water level so a boat could float on top of the berg
between the towers. Lastly, a tabular iceberg has a flat surface
because it usually calves from an ice sheet or shelf glacier. Most
come from Antarctica, though some originate from Greenland
and the ice caps of Arctic islands. These bergs can be many kilometers in length and width—far greater than the powerful ice I
witness in Newfoundland.

Back on land, I head to raucous George Street to be "screeched-
in." The cheeky local tradition begins with a supplication. I
introduce myself and say I'd like to be a Newfoundlander. Next,
I am meant to recite a vow. It is impenetrable to my daft ears,
so I mumble along and jump to the next step, taking a shot of
rum. It goes down easier than the final act: kissing a cod. I am
again grateful the pandemic has changed my plans, since I get
to peck the fish through a mask. I smooch the frozen cod and
quickly order an Iceberg beer. Brewed up the street at Quidi Vidi
Brewery, the smooth lager advertises itself as being "made with
pure 20,000 year old iceberg water." The beer is refreshing, but
I cannot notice anything special about it. Admittedly, I am no
connoisseur, and I've just had a fish in my face. While I drink the
beer, I think of Ed Kean. His sobriquet, Captain Ahab of the Ice,
is unfitting. At the end of *Moby-Dick*, Herman Melville's fictional
captain is drowned in his monomaniacal attempt to subdue the

white whale. In St. John's, the lager in my hand is a sign of Kean's success. He has vanquished icebergs and lived to tell about it.

◆

Besides tour operators who bring admiring visitors close to bergs and fisherman-turned-cowboys who cruise even closer, the North Atlantic Ocean is full of roustabouts, drillers, mechanics, and engineers aboard oil rigs who face the threat of icebergs. They mostly work in the Grand Banks, the series of underwater plateaus southeast of Newfoundland. The area is one of the richest fishing grounds and largest oil fields on earth. It is also one of the most dangerous. Here, the cold Labrador Current meets the warm Gulf Stream and together create extreme fog, just where the steady stream of icebergs traveling south from the Arctic plunge into the wider ocean.

Herman Melville was petrified when he sailed through the Grand Banks as a young sailor. "What is this that we sail through? What palpable obscure? What smoke and reek, as if the whole steaming world were revolving on its axis, as a spit?" Melville writes of the Grand Banks in his 1849 semiautobiographical novel *Redburn: His First Voyage*. A ferocious storm rolls in, and Melville's eponymous narrator reports, "I could hardly stand on my feet, so violent was the motion of the ship." When the sky finally turns blue, the crew passes a drifting wreck. The men aboard had lashed themselves to the taffrail around the ship's stern to avoid being thrown overboard, but had not survived. Now the ship floated on the sea adorned with corpses. Not too much has changed since Melville documented the peril. The RMS *Titanic* sank just south of the deadly patch of Atlantic in 1912, and the commercial fishing vessel the *Andrea Gail* was lost in the Grand Banks during the so-called Perfect Storm of 1991.

Today, oil crews face the same fearsome conditions atop gigantic platforms. The largest in the area, the Hibernia, can house 3,500 people and includes a cafeteria, gym, and swimming pool. Like its neighbors, the Hibernia is a sitting duck for the icebergs that are launched out of Iceberg Alley. C-CORE, a research and development service firm headquartered in St. John's, executes the "ice management" plans at the platform. The company has a long history of dealing with icebergs and even sent a representative to Prince Mohamed Al-Faisal's conference in Iowa back in 1977. I am lucky enough to meet with Freeman Ralph, a vice president at the company, who holds a PhD in ocean and naval architectural engineering. Freeman has silver hair, a scruffy beard, and is wearing a vest over a thick sweater when we talk. As he explains it, he's a true Newfoundlander who "grew up next to icebergs." He tells me that he is willing to travel south, but he will always live in the province. Our conversation is easily sidetracked, since Freeman has a roundabout way of answering questions that often includes fascinating diversions like "I slipped off an iceberg once."

An oil platform has three options when an iceberg is heading toward it: absorb the blow, disconnect from the well it is exploiting and move out of the way, or tow the berg. Hibernia's radar system detects icebergs within eighteen nautical miles, which gives the crew plenty of time to assess the situation and best decide how to respond. The shallow water in which the platform stands provides a natural defense, since the largest icebergs cannot penetrate the eighty-meter-deep water. But smaller bergs can still make it through. According to the engineers of the Hibernia, the impact of a one-million-ton iceberg would leave the platform unscathed and the blow of a six-million-ton berg would only cause repairable damage. The oil companies, though, do not want anything to come into contact with the platform if they can avoid it.

To ensure operations are safe, disconnecting may be necessary, though costly in terms of production losses. This involves depressurizing production wells, flushing flowlines, and preparing to leave location. Freeman cannot recall a disconnection in his twenty years with C-CORE. In 2017 the crew aboard the 270-meter tall *SeaRose*, thirty-one miles from the Hibernia, considered disconnecting when shifting winds and currents unexpectedly hurtled toward it an iceberg 40 meters wide and 60 meters long standing 8 meters above the waterline. As the iceberg approached, Husky Energy—which operates the floating production, storage, and offloading vessel—elected to stand its ground and warned the eighty-four-person crew to brace for impact. In the end, the berg—about the size of a twenty-story building—penetrated the ice exclusion zone but floated by without making impact. The near-miss made headlines in Canada and prompted the Canada-Newfoundland and Labrador Offshore Petroleum Board to suspend Husky's *SeaRose* operations until the energy company demonstrated it could follow its ice management plan. Ralph sighs when we chat about the incident. "Icebergs are a pain in the ass and cost a fortune, but safety first. We are also learning more and more about ice strength and hardness, and what was once thought threatening in 2017, may not be so once we process more of our data and advance our understanding."

When possible, oil companies tow icebergs away from their platforms and vessels. The process is dangerous. At any moment, the iceberg could break up into deadly pieces or overturn and generate explosive waves. That did not stop Ralph from hopping on an iceberg long ago. "I wasn't afraid," the engineer recalls. "I was fueled by adrenaline and had an exit strategy." Actually, Ralph has been on several icebergs for research, but none in the past fifteen years. "It is the riskiest behavior. Liability insurance won't allow it today." Ralph has been involved in many tows, and advises keeping the towing vessels sufficient distance away from the menaces. One could roll or break apart, and the vessel must not be at risk. "Depending on the size of the vessel, of course, there are so many things that could go wrong," he warns.

Towing an iceberg is a bit like wrangling livestock on the open range, but on a much bigger scale. Ralph describes tow vessels as "lassoing" bergs, but clarifies that "we get our rope around the neck of the bronco before we tie the knot." The first step involves dropping hundreds or thousands of feet of rope into the ocean. Ralph uses an ultra-high-strength polypropylene rope since the ice can shred even the thickest towline. The ship then steams around the berg, dragging the rope to encircle it. A mariner catches the other side of the rope with a grapple and closes the snare. All of this while the iceberg, like a stallion, is bucking wildly in some of the most perilous waters of the Atlantic. The vessel crew next attaches a steel towing hawser and weighs down the rope. It is essential to pull the berg at a point below its center of mass, otherwise the force from the tow would cause the berg to overturn; "Like pulling a stump out of the ground," Ralph analogizes. Even so, icebergs easily slip out of towropes. Sometimes, two vessels are deployed. Once close to an iceberg, they shoot a rope line between the ships and move in tandem to chaperone the iceberg away.

It is a slow process. "Maybe the vessel moves at two knots," Ralph explains. He notices my face scrunching in uncertainty and offers an explanation. "If you were out for a walk with your dog at that pace, your dog wouldn't get exercise." It turns out two knots is about the same pace that a turtle moves on land. And it could take hours for the tow ship to build up to that speed. Freeman laughs as he remembers one time he saw an iceberg end up towing a ship. "The berg was big enough so when the current changed, it tugged the little boat where the ice wanted to go." The captain could have increased power, but higher speeds can create vibrations from waves that fracture the ice and cause it to break apart. Increasing the speed could also jolt the berg, which could make it roll and allow it to escape the lasso. The crew would have to recapture the ice and start from the beginning. Depending on the distance required, moving a big berg could take days. Still, in a busy season, crews might tow upward of one hundred icebergs.

Platforms and vessels are equipped with additional tools to move the bergs. Water cannons can shoot thousands of liters of water per minute and can reach hundreds of feet away, depending on the degree the water is shot. These are no squirt guns. In 2021 the Chinese coast guard sprayed Philippine ships in the South China Sea with such power that the latter had to abort their mission and turn around. Blasted at the ice, water cannons can help knock icebergs out of the way. Similarly, crews might deploy underwater propellers to create a jet stream of higher-velocity water to push the ice into a current that will carry it away from the platform. As with towing, the iceberg does not need to be moved far. If they can catch the berg far enough away from the platform, they only need to give it a little nudge. Even a few feet could change its trajectory enough so it sails safely past the platform by the time the waves carry it there.

◈

Simply nudging an iceberg onto a new path requires Herculean effort and engineering savvy. Actually harvesting an iceberg demands brute strength and a lot of guts. Iceberg cowboys in Newfoundland show us that we do not need to surrender to icebergs, despite the warnings of the International Ice Patrol. We can, in fact, subdue them. Recent reports from Russia are even more impressive. In 2016 the Rosneft Oil Company partnered with the Saint Petersburg–based Arctic and Antarctica Research Institute to tow icebergs. According to Rosneft, they were able to successfully lasso a one-million-ton iceberg. Astoundingly, the ships were also able to turn an iceberg ninety degrees from its original trajectory and another completely around. The next year, using the $62 million Novorossiysk icebreaker, the joint venture managed to drag an iceberg through the ice-choked Kara Sea north of Siberia for fifty miles—a far greater distance than any iceberg cowboy has towed an iceberg.

Still, fifty miles will not get an iceberg to New York City, let alone Los Angeles, Dubai, or Cape Town. Ed Kean says he could imagine hauling an iceberg a few dozen miles, maybe even one hundred, "but that would cost you." The Newfoundlander puts an asterisk next to the lesson I learn in Canada. It might be technically feasible to tow icebergs, but the task could be astronomically expensive, especially as the distance increases. Oil companies, including ExxonMobil, are willing to shoulder the cost of roping and redirecting icebergs as part of the cost of operating platforms like the Hibernia, which can produce 220,000 barrels of oil a day. Though, sometimes, that raw revenue is not enough to bother responding to icebergs, as Husky Energy demonstrated. According to the Canada-Newfoundland and Labrador Offshore Petroleum Board, the energy company's decision not to handle the iceberg that threatened *SeaRose* was "economically driven." Commercial motivation will be a key factor in the quest to conquer icebergs.

I leave St. John's and drive the Irish Loop around Newfoundland's Avalon Peninsula. In the picturesque fishing village Brigus, overlooking a harbor pond, I chat with a man while eating fish cakes. When the topic of icebergs comes up, he leans forward and whispers that he has an iceberg in the freezer in his garage. He must be used to impressing tourists with his treasure. I ask how he got the ice. "Oh," he puffs, "I wait till it wash ashore, then I grabbed it." As a newly christened honorary Newfoundlander, I can

attest that this is undoubtedly the safest means of acquiring an iceberg; a strong back and willingness to get one's feet wet is all that's
required for this method. For the businesses that rely on icebergs in
Newfoundland, like Quidi Vidi Brewery and Auk Island Winery,
Ed Kean is willing to do this work on a more ambitious scale. He
estimates that he has harvested more than twenty million liters of
iceberg water in his life. But even Kean is still operating in his own
backyard and, he enlightens, "It hasn't made me rich." Getting an
iceberg to travel around the world is an entirely different matter.

Kean tells me that in his business, "You only have two months
to get the product, and then you have ten months to scheme and
dream." The iceberg cowboy has plans to sell his own iceberg water.
He says he is doing lots of research, working with the university,
thinking about marketing, and creating a business plan. Finding
crews willing to do the work and managing the rising price of
diesel has complicated his dream. Mother Nature complicates
matters, too. "She didn't send any icebergs up this way last year."
Nonetheless, he insists "there's definitely a market for this water,"
repeatedly mentioning New York and Los Angeles as places people
will pay for icebergs. When I ask him what he thinks of large-scale
towing far from home, the most experienced man in the business
titters, "Pie in the sky."

3

Squeezing Profits
from Ice

In 2019 the Royal Canadian Mounted Police got a strange call: water had been stolen. If the report had come from British Columbia's Okanagan Desert, it might have made sense, but the call came from the small community of Port Union in Newfoundland. Located on the Bonavista Peninsula of Canada's easternmost province, the one-thousand-person town clings to a flat, stony clearing where the spruce forest meets the Labrador Sea. Here, the boundary between water and land is porous; Main Street barely sits above sea level and there seem to be as many boats in the bay as there are buildings surrounding it. The pilfered water was different. It was stored in a secure tank in a padlocked warehouse behind a bolted gate. It had been painstakingly harvested by iceberg cowboys and was valued at nearly $10,000. More than ten thousand liters were stolen—enough to fill a tractor trailer tanker. To drain and move that much water, the thieves must have known which specialized gear to use. The Mounties concluded that the heist had been carefully planned.

Fortunately, the water was insured. David Meyers, then CEO of the Canadian Iceberg Vodka Corporation, from which the water was stolen, never expected a theft. He had prepared for potential losses like a leak, since harvesting the ice, in Meyers's words, is such a "dangerous and costly endeavor." Iceberg Vodka, in fact, employed Ed Kean for the arduous and time-sensitive work. "Think about a grape harvest to make wine, you only have one crack at it a year. It's the same thing with icebergs," Meyers explained. The analogy to fine wine is not as absurd as it might initially sound, and Meyers should know. Before joining the Canadian Iceberg Vodka Corporation, he worked as a managing director at Moët Hennessy, where he was responsible for brands like Dom Pérignon. Iceberg water is a similar luxury not just because of the difficult procurement, but also thanks to its content and marketing. According to the advertising materials for Iceberg Vodka, "20,000-year-old icebergs off the coast of Newfoundland contain some of the purest water on earth. It's that same purity that goes into every bottle." The advertising bluster sells, and it apparently makes water valuable enough to pilfer.

According to the Fine Water Society, such statements are not merely hype. The society would have you believe that fine water is indeed the "next wine," since it has a unique terroir that influences its taste. As such, the society insists, water "deserves epicurean attention." Although I am skeptical, I decide to give it just that and plunge into the glamorous world of luxury water. My first stop: the source of the most expensive iceberg water in the world.

◆

In 2016 Jamal Qureshi had a vision. That year, the Norwegian American entrepreneur moved to Svalbard, an archipelago that is officially a part of the Kingdom of Norway but constitutes an "unincorporated internal area" independent of any Norwegian county. The islands lie between seventy-four and eighty-two degrees north, around 800 miles from the North Pole and 1,200 miles from Oslo. Bare black mountains ringed by glaciers dominate the landscape. The air is dry, there is almost no precipitation, and less than 10 percent of the land mass is vegetated. Technically, Svalbard is an Arctic desert in which more polar bears than people reside.

Per the Svalbard Act, anyone, from any country, can live and work in Svalbard—no visa required. There are a few requirements, though. Perhaps most notably, a 2012 law requires anyone traveling outside of settlements to be equipped with "appropriate means of frightening and chasing off polar bears." The governor officially recommends carrying a gun. Even in Longyearbyen, the largest settlement and administrative center of the archipelago, it feels like the Wild West or a fairy tale, or maybe something in between.

Longyearbyen is home to 2,400 people. Some work in the nearby coal mine, but most conduct scientific research or are employed in the tourism industry. Qureshi himself first visited Svalbard on holiday. As the former Wall Street analyst tells it, he missed his wife and collected meltwater for her as a gift. "The pure water would make for a delicious cup of tea," he recalls thinking. Then inspiration hit: iceberg water could be a business. Qureshi relocated to Longyearbyen a few years later to found Svalbarði, a premium iceberg water company. The name is derived from the Old Norse word for "cold coasts" and uses the medieval *ð* or *eth*, which is pronounced like a *th*. Even to Norwegian ears, *Svalbarði* conveys the distant past and conjures an exotic environment.

The first step to creating the business was securing permission to harvest ice. Happily, the governor of Svalbard has wide discretion over matters on the remote outpost. She bears the responsibilities of the executive branch, as well as the chief of police, enforcer of environmental policy, tourism management, and acts as a judge in certain maritime cases. Qureshi, a tall man who looks good wearing an expensive fleece, got the go ahead to collect a few icebergs. Now, he reports, "Business is good."

I visit Svalbard in April, when the sun has finally emerged after a dark winter and the temperature hovers around ten degrees Fahrenheit. Although it takes me a few days to get used to seeing rifles slung over people's backs, I am grateful for the weapons on my first outing beyond Longyearbyen. I have never been to a place where the transition between civilization and wilderness is so fast and so extreme. After driving a snowmobile for fifteen minutes, I am well outside the settlement and beyond the coal mines that first brought people here. On all sides, I am surrounded by snow-covered mountains without a trace of human presence. With a guide, I snake more than fifty miles through narrow valleys until we reach Tempelfjell, or "temple mountain," so named for its resemblance to a gothic cathedral. The massif's nearly vertical walls are skirted by piles of eroded shale and limestone, like flying buttresses on Notre-Dame de Paris. Still, Tempelfjell is far grander than anything humans could build. We turn east toward the Von Post and Tuna glaciers and switch off our engines. Despite my snowsuit, it is bitterly cold in the wind and I ball up my hands inside my gloves. *Humans shouldn't live here*, I think. In fact, until the mining began in the late nineteenth century, they didn't. There is no evidence of Indigenous peoples inhabiting these islands. Tour companies describe this as the "untouched Arctic." It is astonishingly remote; I would be overcome by a sense of dread were the scenery not so spectacular.

This is where icebergs are born. My guide, Vlad, loads additional bullets into his rifle. Ring seals are sunbathing near the glacier front, which means polar bears could be nearby. Northern fulmar glide along the steep cliffs to avoid the Arctic foxes below. The bright sun is reflected off nearly every surface and I am enveloped in silence except the sound of the wind and my boots crunching in the compacted snow. This lunar landscape is dangerous, beautiful, and miraculously pure.

The advertising copy for Svalbarði reads: "Pristine ice, locked up for millennia and fresh as the day it fell as snow, is handpicked during its brief few months of life before it melts away forever in Svalbard's arctic waters. Bottled at 78 degrees north, it becomes Svalbarði." The locals I ask here say they are not interested in the water. They get to enjoy the sublime sights every time they step outside. Qureshi's product is meant instead for an aspirational global clientele. The iceberg water brings a small bit of Svalbard to people around the world.

◆

The West has long romanticized icebergs as exotic marvels of the natural world. Scholars cite this perspective as an example of "Arcticism," an outsider's false vision of the polar north that gets reproduced and naturalized until people forget that it is not the only possible point of view. This discursive practice has resulted in a popular idea of the north as an empty, mysterious, frozen wonderland. The reality, of course, is far more complicated. Yet this Arcticism has been baked into Western culture since its earliest days. Already in 325 B.C.E., the Greek sailor Pytheas of Massalia sailed north of the British Isles and reported a frozen sea, the midnight sun, and the aurora borealis. He claimed there was no proper land, sea, or air, but that the north was a mixture of all three with the consistency of a jellyfish. Icebergs, floating south as enigmatic emissaries from this otherworldly dimension, have been a crucial part of this mythos.

The earliest extant account of an iceberg was recorded around 900 C.E. Known as *The Voyage of Saint Brendan*, the Latin text tells the story of an Irish priest's sixth-century search for the Garden of

Eden. Brendan set out from County Kerry in a small round boat called a currach—made by weaving long, thin pieces of wood into a basket and covering it with animal skin—and, legend has it, spent seven years searching for paradise in the fearsome Atlantic. On his journey, Brendan saw fantastical sights. Mountains that spewed smoke and ash into the air while oozing burning rocks. A sea monster with an enormous tail and the power to shoot water from its head. Today, we recognize Iceland's volcanoes and a whale in Brendan's descriptions. Medieval readers loved to envision the bizarre world Brendan discovered at sea, and *The Voyage of Saint Brendan* became a favorite text in Europe.

One object among the wonders was particularly mysterious. Brendan encountered a "crystal pillar" that pierced the clouds and reached the depths of the ocean. The author of *The Voyage of Saint Brendan* reports that the Irish monk could not make sense of the form, hard as marble, but soft like silver and somehow transparent.

According to the tenth-century account, Brendan sailed his currach through an opening to the center of the pillar, declaring it a "marvel" as the sun shone brilliantly through the glass-like walls. Brendan attached ropes to the iceberg and sat inside it as the sun set. The hollow orb glowed and the monk imagined it to be the eye of God looking over him. Medieval artists struggled to depict the ice. One envisioned a floating castle, and another a skyscraper-like block. In a Latin version of the text from fifteenth-century Germany, the iceberg looks like a refrigerator with tentacles. Unlike volcanoes, which had reference points throughout Europe, the crystal pillar was unrecognizable to Brendan. Eventually, Brendan, awed, simply sails away to the next adventure and the reader is left to contemplate the wondrous iceberg on his or her own.

Ever since, Western audiences have loved to consume icebergs. Nearly everywhere they pop up, icebergs are connected with exotic beauty and sublime adventures. In Jules Verne's 1870 novel, *Twenty Thousand Leagues under the Sea*, Captain Nemo pilots his submarine to the South Pole, where he and his crew encounter breathtaking icebergs. In the frozen forms, Professor Arronox sees "an oriental town with countless minarets and mosques." Then, all of a sudden, an iceberg rolls and the *Nautilus* and her crew are trapped. Despite the immediate danger, the crew cannot help but admire the splendor of the scene. "Words cannot describe the effects produced by our galvanic rays on these huge, whimsically sculpted blocks, whose every angle, ridge, and facet gave off a different glow depending on the nature of the veins running inside the ice." Like Saint Brendan, the crew is enchanted by the play of light on the icy surface, even as it threatens to crush them.

For much of Western history, people have had to make do with secondhand or fictional reports of icebergs, or visual representations of them. Frederic Edwin Church's monumental painting *The Icebergs* is paradigmatic of the attempt to portray the beauty of the colossal ice forms in nineteenth-century painting. Stretching nearly 10 x 6 feet, *The Icebergs* is designed to overwhelm the viewer. Although transparent, the depicted iceberg glows brilliantly, as sapphire blues and polished white ice glimmer in the warm setting sun. Church deliberately frames the left side of the image with ice that stretches out of view, ensuring the iceberg dwarfs the viewer physically and overwhelms her emotionally. Initial reviews suggest viewers had a sublime experience. A critic in the *New-York Daily Tribune* wrote in 1861, "The eye feels the first shock and anticipates in a moment the slow agonies that shall wind . . . about the tender tissues, and the unsuspecting blood-vessels." Originally titled *The North*, Church's painting sought to tantalize viewers with the mysteries of the Arctic. For armchair travelers, works like those by Church and Verne further popularized the idea that icebergs were a rarified treat for those lucky enough to encounter them.

In the twenty-first century, icebergs continue to be depicted as part of a wondrous landscape and are bound with exotic luxury in our imaginations. To that end, in 2010 the French fashion house Chanel dragged an iceberg from Sweden to the Grand Palais in Paris to sell its fantasy fur looks. The 265-ton, 30-foot iceberg stood in the center of the Art Nouveau exhibition hall as models wearing ready-to-wear looks paraded around the melting ice. With each step, the beautiful men and women splashed meltwater that had collected on the runway. Their extravagant looks were framed perfectly by the towering white background and rendered other-worldly by the shocking presence of the ice. "Pure romance," one reviewer declared.

◆

Sitting in front of a gently crackling fire, Jamal Qureshi tells me that he still thinks icebergs are beautiful, he just sometimes looks at them differently now. The businessman does not so much contemplate the icy forms as he does size them up. He tries to select the best source for his next batch of Svalbarði. At first, Svalbard locals thought Qureshi was out of his mind; he recalls them characterizing him as "that crazy American who wants to sell icebergs." Now he has sold four batches of his premium water and needs to bottle more. Qureshi won't share his finances with me, but tells me his margins are good and explains that his business is starting to grow substantially. Luxury retailers like Maison del Gusto in Monaco and Fonsapor in Hong Kong stock the water. Online, Svalbarði retails for €94.95 a bottle. At a high-end restaurant like Ray's and Stark Bar, located inside the Los Angeles County Museum of Art, a bottle goes for $150.

Qureshi admits that the price surprises people. "It is not a cheap process," he justifies. "It involves a lot of effort." Much like Ed Kean, whom Qureshi consulted before undertaking his venture, the Norwegian American heads out to sea to capture icebergs with a crane and net. He is not motoring around the coast of Newfoundland, though. Rather than waiting for icebergs to travel down Iceberg Alley, the Svalbard resident travels to moving glacier fronts located inside fjords north of Tempelfjell. He is after the purest ice, the stuff that has just calved and landed in the fjord. "Freshly born icebergs," Qureshi waxes poetic. These are ostensibly the best bergs because they are from the protected inner part of the glacier and have not collected sediment and been battered by the sea. The blue pieces are better than the clear ones, which could be the result of the ice melting and refreezing. That means the clear ice could contain pollutants from our modern atmosphere. Qureshi hunts for ice with minuscule bubbles that contain only ancient air. He hand-selects icebergs and will put them back if they are not the right quality.

Once Qureshi and his team have their ice, which can take up to a week to collect, they head back to Longyearbyen, where one-ton bergs are heaved into an industrial melter and then collected in a tank. Qureshi bottles on site with help from locals, whom he also hires for his crew at sea. When I express surprise that he decided to build a bottling system so far from mainland Norway, the entrepreneur clarifies, "To support the local economy." Qureshi is modest about the contribution since he employs only a few people. Nevertheless, it is a symbolic investment in the future of the unincorporated Norwegian area, which has committed to cease mining operations by 2023. Qureshi seems to genuinely care about Longyearbyen and the natural world of Svalbard, as well. He eats into his own profits to ensure that Svalbarði is carbon-negative. His company goes beyond the carbon-neutral certification it has earned to support additional CO_2-reducing projects so that "when you enjoy a bottle of Svalbarði, you directly fight global warming."

Qureshi estimates that one hundred kilograms of the North Pole ice cap is saved with every bottle sold.

The final step of the process is affixing a security device to the bottle, known as a proof tag. In addition to a metallic seal that breaks if the cap is twisted to ensure the water has not been tampered with, the tag includes a QR code and unique serial number. Consumers can scan the code and verify the authenticity of their bottle online. The tags are more often used for expensive wines than waters.

Svalbarði has borrowed more than just bottling ideas from the adjacent industry. Qureshi's business plan more closely resembles that deployed by a small, family-owned premium winery than those guiding bottled water brands like Fiji and Perrier. "It is not just chasing volume, volume, volume on thin margins. We have something handcrafted and special so it has different economics to match that." Qureshi also works tirelessly to promote his product, attending trade shows, networking with bigwigs in the world of fine food and beverage, and entering Svalbarði in water-tasting competitions to win notoriety.

Owing to these efforts, the Norwegian iceberg water is available worldwide. The company ships its bottles directly by air from Svalbard and has distributors on four continents, including in London, Sydney, Palm Beach, and Macau. Although the biggest markets are currently the United States and Europe, Qureshi is optimistic that his new push into the Asian market will bring even more success. As Qureshi tells it, his clients "are people who appreciate different waters," and they could be anywhere. While most people drink Svalbardi only on special occasions, a small number drink it regularly.

The Petit Ermitage in West Hollywood, California, is one of the upscale places that serves the iceberg water. The private members' club and hotel has branded itself a magical boutique oasis—the sort of place that greets visitors with Champagne and a fruit platter.

Those lucky enough to gain admission can enjoy artworks by Dalí, Miro, and Rauschenberg on the walls and can luxuriate poolside on the bohemian-inspired rooftop. The romantic outdoor lounge, which aims to be a more relaxed version of the Chateau Marmont or Soho House, is also home to a restaurant and bar. Louise O'Riordan, the club's chief brand officer, tells me that their "water menu is an important point of difference." She guesses theirs is the most extensive water menu in Los Angeles.

Svalbarði is the most expensive item on the menu. That does not mean it gets overlooked. Instead, O'Riordan divulges that some members order it for that reason. At a recent rooftop party for "people in entertainment," as O'Riordan put it, the iceberg water proved a hit. Some of the diners were sober and excited to have a whimsical nonalcoholic option. The price and novelty piqued their curiosity. Then, after having a taste, they were won over and ordered another bottle and another. One celebrity even asked for six bottles to take home to give away as gifts. This is precisely the situation Qureshi has worked tirelessly to create.

◆

Of all the places for Svalbarði to turn up, Berkeley Springs, West Virginia, is among the more surprising. The 650-person town has seen better days. On the drive in, I pass an old mining company and lots of buildings that look abandoned. For the brief moment that US Highway 522 slows down and transforms into Washington Street, the Appalachian hamlet appears quaint. "America's first spa," as the town is known, sprouted up around the mineral springs that bubble up from the ground at this spot. For centuries Indigenous peoples, including the Tuscarora and the Delaware, drank and bathed in the water, which surfaces at a constant seventy-four degrees Fahrenheit and contains sulfates and nitrates thought good for health. In the eighteenth century, George Washington visited and bought property in town. Visitors to Berkeley Springs State Park can still freely splash in the hot springs and marvel at George Washington's in-ground stone-lined bathtub full of the medicinal water. Perhaps best of all, a 1776 law still guarantees the public the right to fill water jugs at the public tap. Thanks to this history, Berkeley Springs is home to the world's largest and longest-running water competition.

I visit in February to track Qureshi's water through the contest. The town is abuzz, and the atmosphere is a strange mix of cosmopolitanism and provincialism. Outside the Country Inn, which hosts the annual water tasting, I spot two pickup trucks emblazoned with the Confederate flag. Inside the nearly one-hundred-year-old Colonial Revival hotel, dozens of waters have been entered from around the globe, traveling from Australia, Japan, Finland, Romania, Costa Rica, and Tajikistan, among other countries. Many people in the audience have come from abroad, too. The majority, however, are residents of the Eastern Panhandle of West Virginia. They are here for the so-called water rush that concludes the contest, when spectators can rush the stage and grab as many bottles as they can carry.

At the center of the banquet hall, bottles of water are arranged in extravagant displays. Green, white, and blue balloons float overhead and blue sashes are twisted around columns, evoking waterfalls. Neat rows of slipcovered chairs face the judges, who are themselves arranged at tiered tables under spotlights. After a training session by the designated "water master," Arthur von Wiesenberger, the judges visually inspect the submitted waters, sniff them, and taste them. Glacial water might be slightly opaque, but the rest should be clear; there should be no aroma; and the taste should be clean, with a light mouthfeel and aftertaste that leaves the drinker thirsty for more.

"Not all waters are created equal," von Wiesenberger pontificates as he tugs on the lapels of his blue sports coat. The middle-aged media personality has been involved with the contest since the 1990s and appears very happy to be talking to the crowd as the master of ceremonies. He helps us understand that the contest is not entirely straightforward. "Not everyone likes the same water," he acknowledges, conceding theirs to be a subjective art. The judges, accordingly, come from across the United States and bring with them their own preferences. Together, they articulate "a result that is reflective of the common opinion," von Wiesenberger decrees.

The whole exercise might sound asinine. The water master admits that the most common question he fields is a variation of "is this really real?" The upshot of the contest, however, is purportedly plain. As von Wiesenberger tells it, Berkeley Springs is "like the Oscars of water" and winning is big business. "The people who win here, the next Monday their phones are ringing off the hook."

Surrounded by the lurid decorations, I am skeptical until I chat with a spectacularly dressed woman from Grenada who is here representing Mount Pure Natural Spring Water. Winning this contest, she explains, would help secure investors in her business. According to the International Bottled Water Association, nearly fifteen billion gallons of water was bottled in 2020. In each of the past ten years, the industry has enjoyed consistent growth. In the United States, bottled water is now the most common consumer beverage, topping carbonated soft drinks, coffee, and alcohol for the largest "volume share of stomach," the Beverage Marketing Corporation reports. The global bottled market is project to reach $505 billion before the end of the decade. An award from the 32nd Annual Berkeley Springs Water Tasting would make a product stand out.

Many of the past winners advertise the accolade. Tešanjski Kiseljak water from Bosnia and Herzegovina, for instance, has incorporated gold and silver Berkeley Springs seals onto its label, showing a row of its past awards and touting "Number one in the world" in its promotional materials. Svalbarði has never topped its category: "bottled, non-carbonated water." It earned a bronze medal in 2017 and placed fourth in 2018. But the Norwegian product has been dominant in the people's choice award for best packaging, winning every year from 2017 until 2021, when it came in a shocking second place.

The audience on this crisp February morning is interested in Svalbarði. Wherever I go in the banquet hall, I overhear people trying to pronounce the name of the exotic water. At the long black table where members of the public can cast paper ballots for their favorite packaging design, I watch spectators pick up the entries to feel the weight in their hands and inspect labels. After one woman reads the Norwegian bottle, I hear her exclaim to herself, "Oh, icebergs!" She turns to me and drawls, "Ain't that extravagant." It seems like the luxury product will fare well in this year's competition.

◆

Since the judges at Berkeley Springs taste each entry blind and cannot share their impressions of the melted Norwegian ice with me, I turn to the world's experts on the matter. The Fine Water Society was founded in 2008 by a Viennese man named Michael Mascha with an exquisite palate and a PhD in anthropology with a specialization in food. After an illness made it impossible for him to safely drink alcohol, the Austrian wine connoisseur began applying his gift to premium water and has become a global ambassador for its uniqueness. The Fine Water Society aims to educate consumers, the media, and food and beverage professionals about "water not just being water." In addition to compiling an encyclopedia of premium bottled waters, Mascha's brainchild organizes tastings and summits, publishes etiquette guides and food-pairing advice, and runs the Fine Water Academy, which offers a Fine Water 101 course as well as a water service certification and water sommelier certification. The former involves several hours of lectures, twenty-eight tests, and takes around two to four weeks to complete. To become a certified water sommelier, students must pass a series of courses taught by water experts in a process that lasts between two and three months.

Most of us have likely noticed that water can taste different, depending on where it originates. Fewer have probably thought about how, exactly, water's terroir influences its unique characteristics. Mascha relies on measurable categories to identify what makes each water special: mineral content, hardness (calcium and magnesium concentration), balance (the level of carbon dioxide), orientation (pH level), and virginalty (nitrogen content). From his perspective, vintage and story matter, too.

The Fine Water Society's analysis of Svalbarði is largely technical. For example, it states that Qureshi's product contains 0.06 mg of potassium and 0.03 mg of nitrate per liter. A number of descriptors are more comprehensible to the untrained: the iceberg water is soft, has exceptionally low minerality, and is very slightly acidic. This light, neutral water, according to the society, would pair nicely with oysters or lobster.

Martin Riese, one of the instructors at the Fine Water Society, is a world-renowned water sommelier, author, and media personality. He designed the water menu for Petit Ermitage, in fact. The native German has bulging biceps, wears stylish clear-framed round spectacles, and is quick to smile. On a call with me from Los Angeles, Riese says he has probably had more Svalbarði than anyone on the planet, even though, in his estimation, it is "not a hydration product." In other words, the iceberg water is not meant to be consumed on a daily basis. It is for special occasions. "You should never drink Svalbarði when you are thirsty," Riese advises. "That is a waste of money." The sommelier describes the iceberg water as "super light" with a "very round mouthfeel" and claims it is "very smooth on the palate." He summarizes: "Beautiful water."

Svalbarði describes itself as "the taste of snow in air." I think it means the water tastes like snowflakes, or how taking a big swig of fresh winter air might taste if it were possible. In any case, the slogan conveys purity and evokes poetic awe. Or, depending on how cynical one is feeling, perhaps the phrase simply induces an eye roll. I had to try Svalbarði myself. I ordered a few bottles online and they quickly made their way from Longyearbyen. Jordan and I invited friends over for the tasting. We didn't dress up or create a special ambience besides pulling out wineglasses, per Riese's recommendation. The "best before" date was still more than two years away, and I had properly chilled the bottle to twelve degrees Celsius. I twisted the wooden cap, broke the silver proof tag, and delicately poured the water. We picked up our glass and hollered "Skål!" in honor of Norway.

The water tasted great. I did not need to complete a specialized course to notice the difference between Svalbarði and the usual water I drink from my tap. The iceberg water was almost like liquid nothingness. Its neutrality was gustatorily striking because it still felt like something in my mouth, more silky than velvety. When I took another sip, I could picture the snow-covered mountains of Longyearbyen and could feel the cold, fresh air of the Arctic. Or, at the very least, I could imagine I was in a posh hotel feeling extravagant.

◆

Back in Berkeley Springs, I chat with a strident woman named Natalie, who lives downtown and makes a point to tell me that she always tries to be friendly to the out-of-towners who visit. She volunteers for the water tasting every year and is stationed at the package voting table today. When I ask who she thinks will win, she rolls her eyes and drops her shoulders like the question has drained all her energy. "Svalbarði," she sighs. "They always win." Perhaps unsurprisingly, Natalie cast her own vote for another bottle. At one point, as he instructs the crowd to vote on best packaging, water master Arthur von Wiesenberger seems to be rooting against Svalbarði, too. He offers sample questions for voters to consider while evaluating the bottles, including "Could the bottle break in the car?" "Is the bottle too heavy to hold?" Of course, Svalbarði is one of the only glass bottles in the contest and weighs far more than any of its competitors. I find myself growing defensive of the Norwegian water. In my opinion, it is the most beautiful bottle on the table.

Svalbarði is held in a crystalline vessel the same size as a wine bottle with gentler curves. Made of extra flint glass, the bottle looks like it could float off the table but is heavy like a paperweight. The bottle's neck subtly widens to accommodate an extra-long cylinder cap made of wood and embossed with a silver snowflake. Apart from a thin aquamarine band near the lip and the reflective proof tag and QR code sealing the bottle, there is little ornamentation. The name SVALBARÐI is printed in black letters along the heel, and a thin two-centimeter paper label repeats the snowflake logo, announces the provenance, states the slogan, and offers serving recommendations. I do not know anyone who would not describe the design as sophisticated. The cardboard sleeve that holds the bottle—or bespoke gift tube, as the company phrases it—is equally tasteful. The thin silver lines of the snowflake design spread across the same pale aquamarine to form a pleasing fractal. It looks like a gift one might pick up from a jeweler.

I find a group of professionals in their twenties from Washington, DC, who are visiting Berkeley Springs for a weekend getaway. They are young and working in an image-obsessed town; certainly they must be charmed by the Arctic product. "How much would you pay for a bottle of water?" I inquire. Turns out, they don't buy water. Elianna tells me that she always has her reusable bottle with her, gently shaking the bottle clutched in her hand. This generation, after all, has come of age when such bottles have become status symbols and sustainability is buzzy. When I ask about places they could not bring their own bottles, Michelle shudders. "At a concert I once paid $8," she replies, the price seared into her memory. When I tell them how much Svalbarði costs, they are incredulous. Collectively, they agree the bottle is intimidating. I give them a little sales pitch, parroting Martin Riese. Any chance they would consume the water like a fine wine? Everyone declines but Charlie. "It is not something I would buy, yet," he pronounces. So there may be hope left for the iceberg water.

Actually, this is exactly how the high-end business model is meant to work. Luxury brands typically do not seek to persuade consumers to purchase their products through traditional means, like discounting or mainstream advertising. Instead, they attract customers by ignoring demand and generating exclusivity for their ostensibly unique products. The heavy bottle on display in West Virginia is meant to be a special product of which not everyone is worthy. Its symbolic value is worth far more than the water it holds. Svalbarði is so protective of its exclusivity, it does not send extra bottles to participate in the water rush that concludes the competition—just enough for the judges to sample. Not just anyone should be able to get their hands on the iceberg water.

On the final day of the contest, the banquet hall begins to fill up. Fifty-year-old men with scraggly beards wearing camo-print hats sit next to women in sparkly blouses and high schoolers sporting varsity jackets. They are busy positioning themselves to best attack the display bottles after the winners are announced. A father and son sitting behind me create a priority list of what to grab. When the water rush finally commences, the crowd pushes forward. One large man scoops up nearly a dozen bottles in his arms. A woman fills up a plastic laundry bag so full, she has to call her husband to help. A few young kids jump excitedly clenching newly won treasures from abroad. No one, though, walks away with iceberg water.

I ask Natalie if she has ever gotten to taste Svalbarði. Once, she replies, when the judges had some left over and offered her a sip. She leans forward and whispers, "Well, one extra bottle happened to jump into my bag last year." "How did you like drinking it?" I probe. Her eyes widen. "Oh, I didn't drink it," Natalie yelps. "It is sitting on a shelf for display. That's a $100 bottle of water! It is just for looking."

In the end, there seems to have been a populist uprising against Svalbarði. Just as the Washingtonians thought the bottle intimidating, many of the locals I spoke with found the water too fancy. It was awarded fourth place for the People's Choice Package Design Award—its worst showing since it began entering the contest. The winning entry, Hawaiian Springs Natural Artesian Water, comes in a pretty standard-looking bottle, remarkable only for its teal color and the large red flower printed on the label. Even though I suspect funny business, since Hawaiian Springs is a prominent sponsor of this year's contest, it still does not explain Svalbarði's underwhelming result. The iceberg water also did not place in its category, losing out to a bottled water from Connecticut.

Most of the entries were not what the Fine Water Society would recognize as "fine waters." According to the society, these are a "luxury segment within the wider category of premium waters." Fine water is "limited in its distribution, unique and rare with a strong aesthetic and emotional content." Further, it has a story "with a touch of dream value."

On this account, Svalbarði is unambiguously succeeding. The iceberg water conjures up the exotic Arctic and offers those lucky enough to afford a sip the chance to experience the taste of snow in air. No wonder, then, that in 2021 the Fine Water Society awarded Qureshi's water a bronze medal in the contest for the best still water with "super low" minerality and bestowed it with a gold medal for Best Glass Design. So long as posh locales like Petit Ermitage continue to stock the flint glass bottles and onetime miners in Appalachian hamlets cannot get their hands on it, the Norwegian brand may just be doing its job.

◆

Martin Riese's mantra is "water is not just water." He means that the life necessity is more complex than most people realize. Terroir, for instance, influences the taste of water. Accordingly, as the sommelier puts it, "Drinking water can take you on a trip around the world." But water is also stranger than most people stop and consider. "We pay for some kinds of water," the German summarizes, "and shit and pee in other kinds water." That is to say: the story behind a water matters, too.

Svalbarði, in my judgment, is worth the expense for its unique taste and even more so for the entertaining experience and story. Drinking an iceberg feels special. For a person like me who grew up in the United States, this is unsurprising, given the long cultural association between icebergs and beauty, opulence, and the divine. It is why adventurers trekking through the Arctic and armchair travelers alike have been dazzled by icebergs and why the natural wonders can help sell luxury goods like Chanel's fantasy furs. Thanks to his savvy business acumen, Qureshi has taken this association one step further and transformed icebergs into a literal treat for those wealthy enough to purchase the rarefied delight.

Other businesses have tried to sell iceberg water without success. Ed Kean is currently struggling to do the same—despite the advice of David Meyers. Qureshi guesses the failed businesses did not price their products appropriately. Harvesting icebergs is more expensive than purifying tap water, say, and thus requires convincing consumers to shell out more money for a product that, to untrained taste buds, may seem identical. Launching a high-end water company is also plainly difficult. The onetime oil market analyst has had to learn about glaciology, shipping and logistics, material science, bottling, and, of course, marketing to make his business successful. Through these efforts, Qureshi has created a strong brand with significant symbolic value. "Right now," he informs me, "the only thing holding Svalbarði back is the need for more capital." When we talk in Svalbard, Qureshi is busy looking for investors.

Depending on one's perspective, the Norwegian American could be seen as a modern-day hero, whose entrepreneurial spirit and shrewd vision has allowed him to achieve success and independence in the competitive business world. Alternatively, he might be seen as a villain exploiting the natural world for personal gain, swooping in before others recognize the value of an unclaimed resource and selling it to those who can pay the premium he demands. Maybe Qureshi is a bit of both.

A number of people have a negative, knee-jerk rejection to selling water at high prices. Advocates of water justice believe that access to clean and sufficient water is a human right. Since the natural resource is necessary for human survival, the argument goes, water is part of a global commons. As such, even though its use is inherently private, the distribution of the resource ought to be as equitable as possible. Consequently, some activists and scholars contend that it is immoral to profit from the sale of water since it belongs to everyone as a public resource. Others argue the private ownership and commercial use of water inevitably weakens public and environmental interests in water. Once water is treated as a commodity, like Qureshi has treated icebergs, rich individuals, companies, or countries will begin to scoop up the resource and sell it for the best price possible. According to this line of thinking, if private greed takes over, it can be difficult to hold on to other values, like protecting the environment or helping those less fortunate, if it cuts into the bottom line. Environmental activists claim that the water business is literally destroying the world. The Swiss water bottler Nestlé has been condemned for drying up surface water resources in California, Coca-Cola has been blamed for stressing water supplies in Rajasthan, and Fiji Water has been criticized for polluting the very environment it uses to market the purity of its product. Commoditizing water is thought to lead to an inequitable distribution of the resource as well, since the market is indifferent to the needs of the poorest people. Fiji Water makes millions of dollars every year, but around 10 percent of the population on the Melanesian island lack a reliable source of clean freshwater. As small as Qureshi's operation in the Arctic is, it may be considered a philosophical and conceptual threat to the humanitarian efforts to get water to those who need it most.

In Longyearbyen, Qureshi told me that he believes "water should be a human right and a delight." He makes an analogy to food: We dine at fancy restaurants, but still believe everyone is entitled to food. Why should water be different? Qureshi, in other words, does not believe that selling iceberg water at exorbitant prices fundamentally changes the character of water such that it will crowd out nonmarket norms like humanitarianism. To most people, though, water is harder to differentiate than food. Because consumers can spot the difference between a steak and lentils, they say they do not think they are denying hungry people a meal by sitting down for a ribeye. Conversely, most consumers cannot distinguish tap water from iceberg water unless they have tried it. So selling water at high prices seems to deprive others of that resource.

Regardless of one's point of view, it is evident that turning icebergs into an emergency water source for impoverished peoples would change the story we tell about icebergs. "It would 100 percent make it harder to sell Svalbarði," Riese predicts. When I ask Qureshi about the possibility, he insists he is too busy running his business to waste time on hypotheticals. The entrepreneur might be right to keep his focus on the present. After all, he has figured out a way to make harvesting icebergs and supplying them to the world profitable. It just comes with a hefty price tag per bottle.

4

Icebergs for the Mass Market

E uropeans were slow to embrace icebergs as a freshwater source. For centuries, drinking melted snow and ice was considered unhealthy. The celebrated German naturalist Georg Forster put it plainly in the eighteenth century: such water is "always known" to cause "swellings in the glands of the throat." Consequently, people were cautioned against the practice. This belief changed when the British explorer James Cook left Cape Town in November 1772 and soon found himself and his crew desperate for water.

Captain Cook first made a name for himself surveying the jagged coast of Newfoundland. There he would have seen many icebergs and gained the skills necessary to navigate treacherous waters around the world. The British government subsequently commissioned Cook to confirm the existence of a *terra australis* at the South Pole. Supplied with two ships, the *Adventure* and *Resolution*, Cook set off on his second circumnavigation of the globe in July 1772. In Cape Town, he loaded up on the essentials—bread and wine—and headed south. Two months later, cold and thirsty, the crew decided to risk approaching the hazardous ice floating around the ships at sixty-one degrees south latitude.

In his journal, Cook describes how his men hopped into boats and searched for loose shards of ice that had splintered off a large iceberg. They rowed as close as possible and broke up the pieces with axes, repeatedly swinging the heavy metal over their heads until smaller fragments could be lifted aboard. The captain pronounces the exertion "a little tedious," and notes the work took hours and the melting still more. Georg Forster, who accompanied Cook on the *Resolution*, was relieved when the crew safely harvested the ice and claims in his journal that the resulting water "was perfectly fresh." In his estimation it even "had a purer taste than any [water] we had on board." The icebergs were a salve for the crew, and Cook followed the laborious procedure throughout his circumnavigation of the hemisphere. The icebergs caused a sensation in England upon the crew's return, when Cook wrote about the water source and other methods of preserving mariner health. Consuming icebergs was such a breakthrough, Cook's paper was awarded the Copley Medal from the Royal Society in 1776.

In his award speech, Royal Society president John Pringle describes water from icebergs as a "wonder of the deep." He expounds that using icebergs for freshwater "had either never been asserted, or had met with little credit." For this reason, Pringle considered the discovery of icebergs' potability the "romance" of Cook's voyage—never mind that the captain had become the first known mariner to venture south of the Antarctic Circle. "Those very . . . floating mountains of ice, among which [Cook] steered his perilous course, and which presented such terrifying prospects of destruction; those, I say, were the very means of his support, by supplying him abundantly with what he most wanted." Thanks to icebergs, Pringle claims, the *Resolution* and *Adventure* were able to maintain their freshwater supplies. Although Pringle fails to recognize northern peoples' use of icebergs as a water source, he was correct that this was a revolutionary discovery for European explorers. It allowed them to continue exploring, to make the world smaller, and claim it as their own.

We can guess that Cook's crew was astonished by the captain's suggestion that they harvest icebergs. Yet they overcame their learned aversion to the water the ice contained, owing to the combination of two potent factors: desperation and bold leadership. Today, many people are facing the same need for freshwater as the crew aboard the *Adventure* and the *Resolution*. To go after icebergs as a solution to this problem, to beat back our knee-jerk reaction to dismiss the idea, we may just need someone with an intrepid spirit and an audacious vision.

◆

Nick Sloane may just be the visionary we need if iceberg towing is to become a reality. His desire to find a solution for the water crisis in Cape Town has convinced him the risk of seeming laughable is worth the reward of collecting these freshwater jewels. Luckily, he is also one of the smartest and bravest people sailing the oceans today.

Sloane was born in the British protectorate Northern Rhodesia, now Zambia, in 1961, and spent his youth sailing on the Indian Ocean. After becoming a master mariner licensed to captain ships of any size anywhere in the world, Sloane turned his attention to salvage operations. When a vessel goes down, whether an oil tanker, ocean liner, or container ship, Sloane knows how to recover it. He has worked across the globe, from Papua New Guinea to New York Harbor, in some of the harshest conditions imaginable: aboard crumbling ships sometimes on fire, often sinking, and spewing chemicals and oil. Sloane is like a maritime Indiana Jones who has rappelled from a helicopter onto a burning ship and battled armed pirates. In 2013 he became famous for salvaging the wrecked Italian cruise ship *Costa Concordia*, which made headlines when it struck a rock off the coast of Tuscany and capsized, killing thirty-three people and causing roughly $2 billion in damage. For his work, Sloane was awarded the prestigious Deutscher Meeres-spreis from the German ocean research organization GEOMAR and Deutsche Bank.

Salvaging a ship like the *Costa Concordia*, three football fields long and over one hundred thousand tons, requires extremely specialized knowledge and careful planning. Over thirty months, Sloane organized more than five hundred people to get the job done. In addition to his own bravery, the salvager is a master of logistics and diplomacy. Still, an unquan-tifiable component is crucial to Sloane's success. According to GEOMAR, it is due to his being a "born optimist." Sloane agrees that his sunny outlook is important but also emphasizes the importance of his intuition.

Now, his gut tells him the future is in icebergs. Sloane is ready to leverage his deep knowledge of the ocean, engineering expertise, and contacts throughout the maritime world to rescue the country he loves and save its people in dire need of freshwater. The first time I heard Sloane speak was in a YouTube video. Coming from someone else, the sentiments he shares would seem hackneyed. "You need to never give up on your dreams," Sloane lilts in his languid accent, "go out and try. Whenever you have an opportunity, take it. And never give up, just keep on going." Out of Sloane's mouth, the words take on weighty importance. His silver locks, coifed in a perfect sideswept part, add a certain gravitas. This is a man who has confronted extraordinary danger and lived to tell his tale. For good reason, Sloane is featured in a lot of inspirational materials. One cannot help but believe in what he says. Neverthe-less, Sloane and his competitors face a few obstacles when it comes to icebergs, including physics.

It does not take a glaciologist to figure out that the biggest challenge of towing an iceberg from Newfoundland to the Canary Islands or Antarctica to South Africa is that the ice will melt before it reaches its destination. Cape Town, for instance, is more than two thousand miles from Antarctica and the water temperature in Table Bay can be fifty degrees Fahrenheit warmer than in the Southern Ocean. To understand the best way to solve the problem, it is helpful to know a little bit about thermodynamics. For that, imagine an ice cube in a glass of water. Why does it melt? The answer is explained by the exchange of energy that occurs.

Ice and water, of course, are the same substance in different states. In a liquid state, water molecules bounce around. Warmer water has more kinetic energy than cooler water—think about boiling water compared to room-temperature water. At lower temperatures, those molecules slow down. Eventually, at thirty-two degrees Fahrenheit, the molecules have lost so much energy they are better able to stick together. They form stable hydrogen bonds, which expand into crystalline forms. The water freezes and becomes solid.

When an ice cube is plopped into a glass of water, energy is transferred. The faster-moving liquid water molecules hit the ice and lose energy. The stable hydrogen bonds, in turn, absorb energy. The water becomes a little cooler and the ice gets a bit warmer. If there is more water than ice, the ice molecules will eventually soak up enough energy that they get excited and break the bonds that hold them together. The ice, in other words, will melt. A similar process happens to the part of the ice cube floating above the water, since room-temperature air contains more kinetic energy than ice. As the molecules in the air collide with the ice, they cause it to melt. Like with water, a greater air temperature will result in more kinetic energy transferred to the ice and thus a faster melt rate.

If you wanted to melt the ice cube in your glass faster, you could try a few tricks. Blowing on the ice would bring additional air molecules into contact with the cube, transferring more energy to the ice. Swirling the water would also help. Since the water nearest the ice will be coldest as the cube melts, the conduction of energy will begin to slow. Stirring the glass would introduce warmer water with more kinetic energy to the ice, accelerating the melt rate compared to letting the ice sit in place.

Conversely, if you wanted to reduce the melt rate, you could add salt to the glass. Because salt water is denser than freshwater, it would sink to the bottom of the glass. This would leave the coldest water—the freshwater melted from the cube—at the top of the glass near the ice, slowing the energy transfer.

The same principles apply to dragging an iceberg to Cape Town or Fujairah. Whether or not an iceberg will make it to its destination depends not just on its size and the distance it will travel but also the water and air temperature, the amount of wind, the ocean currents through which it is pulled, the salinity of the water surrounding it, and the time the berg spends in transport. As roughnecks moving bergs away from oil rigs off the coast of Canada know, going fast is not an option. To minimize the deleterious effects of the energy transfer, there are two main strategies: protect the ice and capture such a large block of ice that it won't matter if some or even most of it melts.

Many people familiar with icebergs, like Ed Kean, Mike Hicks, and Jamal Qureshi, are doubtful the physics can be overcome. Their incredulity is rooted in extensive hands-on experience. They have seen icebergs fall apart mid-tow. They know the amount of work required to wrangle the beasts. They know just how ephemeral these resources are. Such experience, however, may also limit their imagination. Icebergs can be unfathomably big and humankind possesses massive power. Skeptics of long-distance iceberg towing may just need to dream bigger.

◆

Luckily, icebergs have the power to inspire. In the Middle Ages, Saint Brendan saw the divine in their transcendental beauty. The ice was a gift from God, sent for the monk to admire and contemplate. More recently, icebergs have been reinterpreted as an earthly luxury, capable of transporting us to faraway places and selling high-end whims. For the right price, we can even buy a small piece of this splendor to consume for ourselves. The ice itself is responsible for these responses. Whatever it is in our brains that makes a sparkling gem feel rare and valuable is seemingly activated when we see fleeting flashes of icebergs, too.

The blankness of the glistening ice also has a potential downside. It is a perfect surface onto which we can project our fantasies, no matter how absurd. Icebergs thus inspire flights of fancy, from which even the most serious-minded people are not immune. During World War II, British prime minister Winston Churchill, whom King George VI proclaimed "sensible and fairly reasonable," got caught up in an iceberg fantasy when he heard of Project Habakkuk. Devised in 1942 by the inventor Geoffrey Pyke, the project proposed using icebergs, either natural or artificial, as aircraft carriers on which planes could land and shelter. Because all available steel was needed to build ships, tanks, and guns, Pyke decided ice was the best available material, particularly because it naturally floated and it had already been demonstrated—thanks to the efforts of the International Ice Patrol—that the ice was impervious to torpedoes and bombs. An iceberg would just need to be captured, levelled off to provide a runway, then towed to the desired location in the ocean. According to the secret plan, named after the Old Testament prophet Habakkuk, the iceberg aircraft carriers could help fight German U-boats in the Atlantic and expediently position the Allies for the invasion of Japan. Upon learning of the plan from Vice Admiral Lord Louis Mountbatten, Churchill wrote to his chief of staff, General Hastings Ismay, "I attach the greatest importance to the prompt examination of these ideas." He described the advantages of the iceberg aircraft carriers as "so dazzling that that they do not at the moment need to be discussed." Despite Churchill's initial enthusiasm, cooler heads soon prevailed. The above-water surface area of an iceberg was deemed too small for an airfield, and more damning, it was realized that icebergs were liable to turn over suddenly, making them poor places to shelter airplanes. The plan was scuttled by 1943.

Looking back on Project Habakkuk today, we can smugly snicker at the absurdity of the notion. When an audacious idea fails, it becomes laughable with the benefit of hindsight and without the buoying force of hope. Bold ideas involving icebergs are particularly susceptible to seeming extravagant. Perhaps this goes back to the notion that icebergs are divine presences on the planet, or wild animals roaming the maritime frontier; it seems fanciful that we could harness them for earthly human interests. It also does not help that the possibility of towing icebergs has long been treated as a joke.

Although there is evidence that small icebergs were transported from Southern Chile to Peru to supply ice for refrigeration in the nineteenth century, most iceberg towing plans have flopped or were part of a stunt. In 1863 a brief blurb in the *Scientific American* reported that a New England entrepreneur was "fitting up a steamer for the purpose of towing icebergs to India" but dismissed the idea as harebrained. In 1899 the Denver-based *Mining and Industrial Reporter* published a hoax announcement: a company called the Klondike and Cuba Ice Towing and Anti-Yellow Fever Company was engaged in towing icebergs and collecting the gold dust they contained as the ice melted. The prank was supposed to spoof similar Gold Rush announcements, but actually inspired several readers to write in for more information and apply to participate. Other magazines quickly made fun of the people who wrote in. Less than two years after the RMS *Titanic* sank in the Atlantic Ocean, the Northern Berg Ice Company hatched the idea to tow icebergs to cities across the East Coast and placed announcements looking for "tugboat captains and ice dealers who would be interested in towing icebergs into harbors." According to the *Washington Times*, the Northern Berg Ice Company planned to exhibit the ice and then "dynamite the bergs into small pieces for market" in places like Boston, New York, Baltimore, and Philadelphia. The *Boston Globe* compared the endeavor to the exploits of Tom and Jerry, characters from Pierce Egan's popular 1820s series *Life in London*, who became metaphors for rich young men who created chaos wherever they went. The *Washington Times* was more explicit about the bad idea. After explaining that "no names of interested capitalists have as yet been made public," the *Times* continued, "the advertisements this morning came as a surprise to ice dealers, who say that the scheme is not practical." There is no evidence that the Northern Berg Ice Company ever launched, but readers at least enjoyed a good chuckle—all this well before Dick Smith tugged his foam-covered barge through the Sydney Heads.

Much of the public is enjoying a similar laugh about iceberg towing today. This is in part due to the hype surrounding some of the best-known schemes and the resulting letdown when they fail to materialize. In 2017 the United Arab Emirates–based company the National Advisor Bureau Limited first announced its "UAE Iceberg Project" and has been subject to derision ever since. Led by the inventor Abdulla Alshehhi, the project and its bold claims to drag an iceberg to Fujairah on the Gulf of Oman was covered by the media around the world. News from Dubai explained that the project would make the UAE "the first desert country to offer glacial tourism," since, in addition to providing freshwater to people in the Emirates more affordably than desalination, Alshehhi claimed the iceberg would be a tourist draw. It would further alter the weather to bring more rain to the region, benefitting local agriculture as well. To drum up excitement, the UAE Iceberg Project released splashy promotional videos visualizing the plan. They show tugboats circling a tabular berg in the Southern Ocean and dragging it to the sandy beaches of Fujairah. Animated spectators gather under palm trees to marvel at the enormous block of ice, on top of which penguins and polar bears frolic—never mind that the species occupy different poles. Excitement, not science, is at the forefront of the presentation. In the end, the video shows the desert transformed into an endless green oasis. In 2019 it was reported that the test run would cost between $60 and $80 million and the price tag for a full tow to the UAE would come in between $100 and $150 million. The same year, the UAE Iceberg Project announced plans to tow a berg. It never happened. The years 2020 and 2021 also passed without a single penguin, polar bear, bergy bit, or growler being towed to the Emirates city. Perhaps fairly, the UAE Iceberg Project looks like another stunt in the long history of harebrained dreams about icebergs.

◆

Nick Sloane's plan seems different. He has assembled a top-notch team of glaciologists, engineers, and oceanographers, including Dr. Olav Orheim. The Norwegian iceberg expert has been looking for solutions to iceberg towing ever since he attended the 1977 conference in Ames, Iowa. Likewise, the French engineer Georges Mougin, who served as the CEO of Prince Mohamed's Iceberg Towing International, is a happy part of Sloane's team. And Sloane, of course, can also rely on his own seafaring expertise to realize his effort, which he has named the Southern Ice Project.

Sloane is focused exclusively on Antarctica. The bergs there are the right size, shape, and distance from South Africa in Sloane's estimation. "To make it economically feasible," Sloane explained in 2019, "the iceberg would have to be big." Somewhere in the neighborhood of 125 million tons and about 1,000 meters from end to end, or about as long as Table Mountain is tall. Luckily, icebergs in the Southern Ocean can be gigantic, like Iceberg B-15, which stretched 295 kilometers by 37 kilometers, or the same size as Jamaica. The behemoth calved from Antarctica's Ross Ice Shelf in 2000 and took more than two decades to break apart into pieces too small to track. In the ice-free part of the Southern Ocean, typical icebergs have a diameter of 300–500 meters, but there are still plenty that exceed 1 kilometer. Antarctic icebergs also tend to be tabular, which are more stable than the bergs that calve in the Arctic. Ships towing these icebergs would have to disconnect less frequently, saving time and money. Nevertheless, some disconnections can be expected, since these icebergs are found in some of the most dangerous water in the world. Around Antarctica, waves can swell as high as a five-story building and winds can reach 80 mph. To make matters worse, thousands of hull-slicing icebergs and shrapnel-like bergy bits float in the water like mines. This is where Harriet M'Intyre and her children died aboard the *Guiding Star* all those years ago.

Sloane has some advantages over nineteenth-century captains. To start, the salvager will use satellite data like that created by the International Ice Patrol to find an iceberg that is the right size and shape. The team is looking for a berg near Gough Island, midway between Antarctica and South Africa. Using radar and sonar, the mariners will then inspect the ice for structural defects. If it passes muster, they will lasso the berg using a giant net—estimated cost $25 million—and rope and tugboats, like the oil companies supported by C-CORE off the coast of Newfoundland.

Next, two tankers will pick up the net and slowly reach a maximum speed of 1 mph. The team does not intend to tug the berg the entire 1,600-mile journey to Cape Town. Instead, they will use the Antarctic Circumpolar Current to drag the ice east, then, at the right moment, pull the ice into the north-flowing Benguela Current toward its final destination. Sloane estimates the journey will take just shy of three months. Even if storms cause more erosion than anticipated, the team expects to have a good percentage of the ice intact by the time they reach Mother City.

Mougin has been working on this problem for decades. Using algorithmic simulations of how an iceberg might crack and break apart, mapping heat exchanges between the iceberg and the sea and the air, using meteorological and oceanographic data from satellites to track temperature, weather conditions, wind loads, and currents, Mougin has shown that the feat of towing a berg far distances is both mathematically and physically possible. Over a decade ago, the French engineer calculated that a multimillion-ton iceberg could be towed from Newfoundland to the Canary Islands in 141 days using just one tugboat and would only lose 38 percent of its mass. Now, the focus is on Antarctic bergs, and the expected results are even better.

When I chat with Dr. Olav Orheim to discuss the feasibility of Sloane's plan, he is quick to move beyond the modeling. "It is all established," he tells me by phone from Norway. I understand where his confidence comes from. Dr. Orheim is among the world experts on ice and probably has done more fieldwork atop icebergs than anyone on the planet. "In the 1970s," Dr. Orheim tells me, "we didn't know anything about icebergs compared to today." Besides better information about the ice, we have more advanced satellite data and a more sophisticated knowledge of the ocean. The required gear has been tested now, too. Strictly speaking, no new technologies are required to execute Sloane's idea. The same equipment used by folks like Ed Kean and corporations like ExxonMobil can be deployed in the Southern Ocean, just on a bigger scale. The theoretical possibility of towing icebergs from the poles to places in need of water, from the glaciologist's perspective, is settled.

Only some practical considerations remain. Dr. Orheim points out that we still have a poor understanding of the internal strength of icebergs, which is important if you're considering dragging them through rough waters for a great distance. Ocean eddies—circular currents that wander around the ocean and can stretch hundreds of kilometers—could also lead to dangerous breakups and launch the iceberg in unexpected directions. But Dr. Orheim is optimistic. "You can get help from the eddy, if you know where it is and can properly position the berg in it." The trouble is that it is hard to know where eddies may appear, how long they will last, and how fast they will move. I begin to question the strength of the model with all of these unknowns, and Dr. Orheim cuts me off. "There are many issues and questions," he admits. "We are trying something that has never been done before. Finding an iceberg of the right size and internal strength may be difficult." There are always going to be some unanswered questions, it seems. "The biggest question," he tells me, "is who is willing to *try* something?"

The most speculative work actually begins once the ice is in its new location. Sloane suggests mooring the iceberg many miles from land in comparatively cooler waters and wrapping the underside of the ice in a heavy geotextile skirt—estimated cost $22 million—to help prevent further melting. One possibility involves anchoring the ice in an old submarine channel just north of Cape Town. Machines will be brought out to the ice to excavate and pump the resulting slurry into container ships that will ferry the freshwater back to land, where they will be piped into the municipal reservoirs. The team figures the iceberg could produce around forty million gallons of useable water daily for an entire year.

Sloane claims he was ready in 2019 to tow an iceberg to Cape Town, yet no berg arrived. The project faltered because of the hesitation of the local government, which would not agree to help pay for the water. Bloomberg reported a total cost of more than $200 million. Other estimates put the figure at $100 million to tow the iceberg—assuming it did not break apart in transit—and more than $50 million to harvest the ice on site. The idealistic ship salvager freely admits that the cost would be considerably more than what the city currently pays for the delivery of surface water. Still, Sloane believes in his vision; he has reportedly invested $100,000 of his own money into the project. Nevertheless, Cape Town officials dismissed the effort as too uncertain and too costly. Even if an iceberg could be towed, they had a difficult time envisioning how it could be injected into the existing water supply. Dr. Orheim confesses this is the advantage of the UAE Iceberg Project: the entrepreneur and the government are more willing to take a risk and outlay the capital necessary to capture the freshwater.

◆

Luckily, we have additional visionaries. POLEWATER, a German company headed by Timm Schwarzer and Heiner Schwer, has been working on iceberg towing for more than a decade—longer than Sloane, Schwarzer points out. The business people have a different plan to bring the frozen freshwater to the western coast of Africa, the Caribbean, and beyond. Remarkably, they want to give away water for free to those in dire need.

The team plans to use satellites to identify favorable icebergs, but they will not tow them to locations that need water. Instead, they will take them to regions where it is possible to harvest the water. The strategy is to "harness nature to do the work for us," Schwer explains to me over the phone from Germany, outlining the same plan to use the Antarctic Circumpolar and Benguela Currents. The POLEWATER team has partnered with the Fraunhofer Society for the Advancement of Applied Research to develop new technologies that will make the transport of melted water more efficient and the anchoring of towed icebergs more environmentally friendly. Writing from Germany, Schwarzer clarifies, "The difficulty is not finding or transporting ice, it is anchoring and storing the water. How do you stop the iceberg? If we don't stop the iceberg, it would either disrupt shipping or run aground off a coast."

Suction buckets, which have been tested on offshore wind farms, will be used to keep the iceberg in place rather than drilling into the seabed. Once the iceberg is in a fixed positios, POLEWATER will rely on the sun to do more of the work. Since water will pool on the top of a tabular iceberg as it melts, the team designed a mobile, floating water station that pumps, filters, and analyzes the quality of the water as it is pumped into bags. Unlike the operations in Canada and Norway that require hacking apart the ice, feeding it into industrial melters, and then storing the liquid water on land, POLEWATER will only have to pump the melted water into easily transportable bags. The team ruled out large tankers as environmentally and economically unfriendly, so they developed the special bag to hold the water without any energy expenditure. Schwer assures me "all of the technical issues are solved." They know where to get all of the equipment "from the oil and gas industry" and already have an offer from a third party to fabricate the water bags. The German company hasn't tested the plan yet because it does not have the financing.

In Schwer's estimation, his team needs millions of dollars to get the operation up and running. POLEWATER is calculatedly registered as a limited liability corporation in Germany and not as a nonprofit. In order to ensure a continuous and secure supply in remote regions of the world, technical and operational stability is required, Schwarzer explains to me by e-mail in German. "This requires financial stability and leads us to a classically functioning business enterprise. We don't want to become dependent on donations." Schwer is a bit dismissive of Sloane whenever I bring him up. "He seems mostly interested in making money," Schwer scoffs. As he describes it, "Nick told South Africa, bring me $400 million and I will bring you an iceberg. But it is more complicated than that, especially if you want to help people." Sloane's own experience suggests it can be difficult to convince governments to gamble paying on something that seems extravagant. By opening up the project to investors, POLEWATER hopes to secure financing to move forward with a tow. "People will only invest large sums in a risky project if they get some sort of reward, too." Schwer and Schwarzer believe this is the best way to achieve the humanitarian aims.

POLEWATER plans to sell bottled water to subsidize the charitable arm of the business. "Let the rich people pay for the water and those who need it get it for free." The plan is built into the marketing for the premium water. When Schwer describes why the plan to sell expensive water will work, he uses the familiar vocabulary of Arcticism and evokes the well-established ideas of purity and exclusivity associated with icebergs. The German team also includes a water sommelier certified by the Fine Water Society, who will be in charge of the bottled water. Schwarzer notes, though, that "we will not specifically market ourselves as premium water. In terms of purity, it is a very high quality water. In terms of production, it's more of an adventure water."

POLEWATER's ultimate aim is to support regions in need of drinking water with a stable source of water without a break and without delay. That won't happen instantly, so the company has devised a multi-part plan. The first involves the technical implementation and financing of the company through the sale of bottled water. In this phase, POLEWATER hopes to deliver water to an area struck by a natural disaster on a short-term basis while they gain experience and stabilize the business to prepare for large-volume water supply. Financing of ad-hoc aid is already built into POLEWATER's business plan with the goal of being able to provide fast, unbureaucratic help. Schwer is focused on disasters. "An earthquake or a deadly drought." The business would absorb the cost of the tugboat and towing the water bags. "We could bring water to Haiti in five to seven days after a disaster," Schwer prognosticates. "And the international press would be advertisement for the premium water that people could buy to support the charity." It seems logical enough.

Once the processes pass the stress test, the German company will offer water to countries for a fee. As Schwarzer details, "Countries can tell their need. With the information about quantity, time, and period, we can automatically calculate when an iceberg has to be transported, in what size and number, and via which route." In this instance, Schwarzer tells me, POLEWATER is a "beverage retailer." A state with an anticipated water need must not make any preliminary investments and must not co-finance years of planning. It can simply order water and POLEWATER will deliver it.

My discussion with Schwer gets vaguer when I ask how much it would cost a place like Cape Town. The Germans can only theoretically calculate how much of the ice will arrive and how much water can be extracted from it, so they cannot say for how much they would sell the water or if it would make sense for them to enter into long-term supply contracts. Prudently their business plan is conservative and assumes only a small amount of water will be harvested from the first attempt. "We would be satisfied to get 10%," Schwarzer expounds, "With fantasies that are too big, we will not be able to support the world." "We just need to tug one iceberg," Schwer tells me, "then we will know more." In other words, POLEWATER is facing some of the same uncertainties as the other projects. The biggest question for Schwer and Schwarzer, like for Sloane, is not whether towing icebergs is technologically feasible, but whether they can secure investors to help this vision become a reality.

◆

In the Old Testament, Habakkuk records a message he received after crying for help and feeling ignored. God tells the Hebrew prophet that he will "be utterly amazed" by what will come. "For I am going to do something in your days that you would not believe, even if you were told." In the end, although he cannot make sense of it, Habakkuk decides to believe in the prophecy. He has faith that something unbelievable might save him. Pyke, Mountbatten, and Churchill were ready to see amazing things in icebergs during World War II, like Saint Brendan before them. Although Project Habakkuk might have been a failure, icebergs might still spare us from disaster. Alshehhi, Sloane, and Schwer may be prophets of a sort today.

These visionaries know that many people scoff when they discuss dragging icebergs from the poles to warmer climes. They know it will be difficult and expensive. But they also believe it is time for another revolution when it comes to the way we think about our drinking water. Admittedly, this process takes time. During the 1912 congressional investigation of the sinking of the RMS *Titanic*, Senator William Alden asked Fourth Officer Joseph Groves Boxhall, "What are [icebergs] composed of, if you know?" The inanity of the question subjected Alden to considerable ridicule among the educated at the time. However, historians have since suggested that such a simple question was designed to compel the officer to offer information the general public may have been wanting. In other words, most people at the start of the twentieth century likely did not know that icebergs were comprised of freshwater. If this is the case, Georg Forster and Captain Cook's "discovery" of the potability of the ancient ice was slow-spreading news.

The most common response I get when I tell people about this book is confusion. *Icebergs are made of freshwater? I never thought of that.* The second most common response is laughter. The idea of dragging a block of ice across the ocean to warmer waters seems absurd. More than fifty years after Prince Mohamed Al-Faisal visited Iowa to explore the feasibility of towing, have we learned enough and is the public ready to accept the results?

Fortunately, we do not need to have blind faith like Habakkuk that some divine force will perform an incredible feat for our benefit. We have all the necessary technologies—already used by organizations like the International Ice Patrol—to identify and track icebergs. And iceberg cowboys have already tested the equipment that will be used to tow the frozen behemoths to cities in need. We also do not have to put our trust solely in a handful of idealistic entrepreneurs, since the possibility of towing icebergs has been modeled by third parties without financial stakes in the effort. Dr. Alan Condron, a research scientist and climate modeler at Woods Hole Oceanographic Institute, has simulated long-distance towing from Antarctica to Cape Town to show that the melting that occurs in transit will still leave enough freshwater to make the endeavor worth the effort. A comparatively small iceberg at 2,000 feet long and 650 feet thick at the time of capture could supply Cape Town water for a year. Condron made this calculation assuming no insulating material would be wrapped around the iceberg to slow the melt rate. But, if left unprotected at its destination, an iceberg would melt within days or weeks in the warmer waters off the coast of Africa. Accordingly, the meltwater would need to be collected and stored very quickly—for example, in a water bag. Or a larger iceberg would work. Condron estimates the berg would need to be at least five kilometers long at the time of capture to supply Mother City with water for a year. Or the berg would need to be protected. Propitiously, an English patent was granted in 2022 for flexible, heat-insulated "Iceberg Reservoirs" that will prevent ice from melting. Alshehhi is the proud owner. Despite these calculations, some faith will still be needed. "At some point, you can throw all the modeling you have at it," Condron tells me over a Zoom call, "but you just need someone to go out and do it."

Convincing people about the math and the thermodynamics is only part of the equation. The other task is persuading potential investors to open their wallets. POLEWATER, for example, says they need €8 million initially. The UAE Iceberg Project is currently looking to raise $9 million, but this is just for the first round of funding. Given such large sums, it is undoubtedly beneficial that several groups are trying to harvest icebergs and are pursuing different business models, whether relying on private individuals, corporations, governments, or a combination of all three.

Schwer chuckles when I ask him about Jamal Qureshi and the high price of Svalbarði. "Good for him, he can do what he wants." It won't change POLEWATER's plan. Dr. Orheim responds similarly. "Why would anyone pay so much for water?" For his part, Qureshi is skeptical large-scale towing will ever happen. "Let them try if they want." The groups are willing to let the free market sort out this matter. Their tunes change, though, when I bring up the possibility of a corporation like Nestlé or Coca-Cola getting into the game. With deep pockets, such companies could more deftly move into the ocean and tap into this resource. They might also be focused exclusively on profits. "It would not be good for the welfare of the world," Schwer portends, let alone for businesses like Svalbarði or POLEWATER. For many people, it is easy to imagine conglomerates as the enemy of the public good when it comes to iceberg harvesting. However, if we expand our conception of the planet beyond the human, Schwer's concern for the welfare of the world might implicate his own effort as well. Despite the noble goal of supplying freshwater to those in need, dragging ice around the globe may have devastating effects on the planet. Even if we can overcome the technological and thermodynamic challenges and raise the funds necessary to achieve the feat of towing icebergs, should we?

5

Bending the Global Arc

In the first pages of Herman Melville's classic novel *Moby-Dick; or, The Whale*, the narrator Ishmael describes a fantasy: icebergs in the Moluccas, an Indonesian archipelago on the equator once known as the Spice Islands. From Melville's perspective, icebergs represented an exotic object confined to faraway, frozen places. Only in a dream could an iceberg be found floating in warm waters. Melville would likely have had a hard time comprehending an idea like that pitched by Nick Sloane, Abdulla Alshehhi, and Heiner Schwer, even if he had learned of Captain James Cook's groundbreaking discovery of the potability of iceberg water. The freezing giants look different today, and so does the world.

Towing an iceberg across the globe is a bit like time travel. Whether they are consumed as an emergency freshwater source or as a luxury product, icebergs are valued for their purity. Drinking one is like taking a sip of the distant past, before the planet was filled with human-made pollution. But it is unclear into which era we are dragging them.

Antarctica began to ice in the Eocene epoch some forty-five million years ago, and the Greenland ice sheet first formed eleven million years ago during the Miocene epoch. Geologists divide time according to rock layers and the history they record.

Each resulting epoch is chemically or biologically distinct. For instance, fossils of dinosaurs like the stegosaurus and bronto-saurus are found in strata dating from 150 million years ago, or the Late Jurassic epoch, but are not found in layers from other epochs. The Holocene epoch, which began about twelve thousand years ago, witnessed the extinction of wooly mammoths and saber-toothed cats. Now, many scientists believe we are living in a new geological epoch dubbed the Anthropocene.

Named for the Greek word for human, *anthropos*, the Anthro-pocene marks the impact of *Homo sapiens* on the planet. Record carbon dioxide emissions, ocean acidification, and widespread habitat destruction have fundamentally altered Earth's chemical, biological, and physical systems. In the future, scientists will be able to dig back down to our present layer and see that it is different than what came before. Today, no part of the planet remains untouched by human activity. To give just one example: microplastics have been found in the Yangtze River in China, Argentina's Patagonian Desert, and on Antarctica. The small plastic debris has even been found atop Mount Everest and deep in the Mariana Trench.

Given the widespread influence of humans today, it is difficult to claim that anything that happens in nature is "natural." This is unmistakably evident in Greenland, where the ice sheet is melting faster than in any period in the past twelve thousand years. Before the Industrial Revolution, the ice waxed and waned as the planet warmed and cooled. Increased carbon emissions and rising global temperatures, however, have changed that pattern. The ice sheet is currently shrinking faster than it can amass new ice. Between 2003 and 2019, it lost an average of 255 billion tons a year. In 2019 alone, the ice shrank by 532 billion tons. Glaciologists believe the tipping point has been reached and the sheet will sustain mass loss for the foreseeable future.

Besides the physical problems such a reality creates, like rising sea levels, the Anthropocene poses new conceptual challenges. In this epoch, when human-induced global warming is triggering iceberg calving at unprecedented rates, it is hard to say if harvesting an iceberg is more or less natural than letting it stay put. We cannot simply claim that it is environmentally wrong to tow an iceberg. Instead, we must ask how removing icebergs might impact the polar regions and the planet overall, since the act may both exacerbate and ameliorate the damage we have caused.

Could moving an iceberg from Antarctica to Cape Town result in a beached orca in Mumbai? The question sounds like a distortion of the butterfly effect—the chaos theory hypothesis that the flap of a butterfly's wings in Brazil could set off a chain reaction that results in a tornado in Texas. In reality, the iceberg inquiry is a matter of conservation biology and ecosystem collapse. In elementaty school, lessons about ecosystems are often accompanied by drawings of idealized landscapes. Some are dominated by lush forests populated with predators that feed from fish-filled lakes in perfect balance. Others depict deserts with cacti, vultures, and subterranean insects living in harmony. Different biomes, or geographical areas with particular climates, sustain specific communities of plants and animals and thus different ecosystems. The goal of these illustrations is to teach how energy flows through an ecosystem and the ways nutrients cycle within it. Most students are too busy learning the parts of these complex networks to think about what might happen if a foreign component makes a surprise appearance. Biologists call this an "invasion."

Even without direct human intervention, icebergs are not confined to polar biomes. Masses from Greenland have been spotted as far south as the Azores, the Portuguese-controlled Atlantic archipelago that sits at the same latitude as Sicily, and icebergs from Antarctica have ventured as far north as Viedma, Argentina. These, however, are the exception. Most bergs are grounded after calving and sit in the same location for years. Few make it beyond the poles.

What happens when an iceberg is towed across the ocean and invades a new ecosystem? Moving an abiotic component like an iceberg into a different ecosystem would be more complex than simply adding additional freshwater to the locale. The disappearance of the frozen mountain would affect the ocean from which it was captured, too, and everything along the way to its new home. If an iceberg were to invade a new ecosystem, the consequences could ripple across the globe, affecting plants, animals, the currents, and even the weather.

In Abdulla Alshehhi's garish promotional video for the UAE Iceberg Project, a simulated iceberg is plunked near a beach in Fujairah, even further from the South Pole than the Moluccas. In the end, the viewer is left to gawk—together with a simulated crowd—at frolicsome polar bears and penguins atop the sparkling ice. Although in real life, of course, the species' habitats do not overlap, the spirit of the presentation could be right. When it comes to iceberg towing, we have gone far beyond the realm of what is expected. Alshehhi's ecological mistake is, in its own way, telling. Iceberg harvesting will muddle existing ecosystems in novel ways.

The Portuguese sailors who named the Cape of Good Hope mixed up the world in a new way when they realized they could bring Indian spices by sea back to Lisbon around the southern tip of Africa. In the twenty-first century, we have even more astounding ways of connecting far-flung parts of the globe. We can make Ishmael's fantasy a reality. Whether we should be as optimistic about the current endeavor as the fifteenth-century mariners were about their newfound trade route is an altogether different issue. If we bend the global arc and introduce icebergs to new places, there might be broad consequences for the planet. To that end, I have tracked down brilliant scientists who have considered the possibility of iceberg harvesting to learn more about the environmental matters we ought to consider if we undertake this pursuit.

◆

Dr. Ellen Stone Mosley-Thompson is one of the world's foremost experts on ice. She is an elected member of the American Academy of Arts and Sciences, part of an elite group of intellectuals and artists that includes Thomas Jefferson, John James Audubon, Robert Oppenheimer, Willa Cather, Duke Ellington, and more than 250 Nobel Prize winners. Dr. Mosley-Thompson has not won a Nobel Prize herself because the committee does not award one for her discipline: paleoclimatology, the study of past climates. If they did, there is no doubt she should win it. Dr. Mosley-Thompson helped transform paleoclimatology into a full-fledged scientific subject, traveling the world to collect ice core samples and analyzing them to recreate the climatological record of the planet. Maybe better than winning a Nobel Prize, part of Antarctica is named after her. The Mosley-Thompson Cirques are steep-walled, curved valleys that scoop into the 8,650-foot Colwell Massif, a mountain range about the height of Yosemite's Half Dome.

Dr. Mosley-Thompson has led nine expeditions to Antarctica to retrieve ice cores that unlock the secrets to climate change and glacier retreat. By drilling into the ice, sometimes a mile deep, she is able to piece together an archaeological record of the ancient past: carbon dioxide levels, what kind of dust floated through the air, past air temperatures. In addition to publishing the results of these studies in hundreds of scientific articles, Dr. Mosley-Thompson has testified before Congress on the dangers of climate change, consulted on the Academy Award–winning documentary

An Inconvenient Truth, and continues now, at seventy years old, to educate the public about the global crises she has witnessed up close.

Today Dr. Mosley-Thompson looks like she is still ready to climb the Himalayas or fly to Antarctica to collect more ice. She is a small, wiry woman with platinum hair and light blue eyes that seem to glow. Although she has not lived in her home state since her college days, Dr. Mosley-Thompson still speaks with a West Virginia accent, drawling out words like *alpine* with two distinct and stressed syllables. The "ice hunter" and "Indiana Jones of glaciers," as Dr. Mosley-Thompson is known among academics, grew up in Charlestown, in the Eastern Panhandle of West Virginia just an hour southeast of Berkeley Springs, where the international water contest takes place. Dr. Mosley-Thompson was not thinking about water at the time. As an undergraduate at Marshall University, she was the only female physics major among her classmates. Her attention first turned to glaciers in graduate school.

"Make sure you write that I am skeptical of iceberg towing," Dr. Mosley-Thompson instructs me. To write this book, I have talked to dozens of scientists. They are uniformly dubious. I have also noticed another commonality among the glaciologists, oceanographers, and climate scientists I interviewed. Many struggle with the hypothetical questions I pose. They want to address the reality of the situation. *Oh, that will never happen*, they reply when I ask about the environmental consequences of iceberg towing. This response gets them out of answering the question, which I have come to learn does not have a particularly good answer. Luckily, Dr. Mosley-Thompson is completely game to indulge my hypotheticals about icebergs. She also connected me with other "big-picture thinkers," as she calls them. They are, collectively, some of the brightest people I have ever encountered, and have studied countless miles of the Arctic and the Antarctic using satellites, computers, and their own eyes and feet.

◆

The area around the North Pole might seem desolate and harsh to the untrained eye, but it is far from lifeless. Besides people, more than five thousand animal species inhabit the Arctic, from ferocious 1,500-pound polar bears to the tiny woolly bear moth that can survive at negative ninety degrees Fahrenheit. Arctic marine ecosystems are especially life-abundant. Over two thousand species of algae grow in the waters, and countless microbes recycle nutrients within the Arctic seawater, sea ice, and sediment. The Southern Ocean is similarly teeming with life. Penguins cavort atop the ice, whales splash in the chilly waters, and tube-feeding worms and stationary filter feeders thrive underneath the ice sheet.

It is impossible to talk about a singular polar, or even Arctic or Antarctic, marine ecosystem. For instance, the North Water Polynya, an area of open water that never freezes in Baffin Bay, is home to more than eight thousand narwhals. Five hundred miles south, bearded seals whelp in Disko Bay precisely because ice appears there in winter. The three words I encounter most frequently when reading about Arctic and Antarctic ecosystems are *rich, fragile,* and *unique.*

As an iceberg melts, it creates a small ecosystem of its own, though. This allows us to roughly surmise the environmental impact of iceberg harvesting in both the Arctic and the Antarctic. After calving, an iceberg freshens the surrounding ocean water and releases minerals long trapped inside of it. The denser, saltier water around the iceberg sinks and causes water to circulate, creating an upwelling current that further brings nutrient-dense water from the bottom to the surface. A web of life is then built around, underneath, and above the iceberg.

In the Southern Ocean, scientists have reconstructed various components of this ecosystem. The iceberg releases inorganic nutrients like nitrates, phosphates, iron, and sulfur. Single-celled algae like diatoms use these minerals and their own photosynthesizing power to generate energy and grow. Krill then feed on these phytoplankton and make the cracks and crevices of the iceberg their home. The scientists found baby icefish, comb jellies, and segmented worms swimming among the phytoplankton, too.

An iceberg's life-giving reach extends well beyond its surface. A ring of phytoplankton surrounds the ice as nutrients are released into the water. Depending on the size of the berg, the radius of enrichment can stretch for kilometers. This attracts life to the ice above and below the waves. Seabirds, for instance, eat the organisms sustained by the iceberg. Among others, fulmars and Cape petrels are two to six times more abundant within half a kilometer of an iceberg. Icebergs, in other words, influence the number and distribution of seabirds over the ocean. Larger animals rely on icebergs, too. Top predators, including chinstrap penguins and Antarctic fur seals, are known to associate with icebergs. Baleen whales also subsist on krill, which are supported by the frozen freshwater blocks.

The enhanced phytoplankton biomass and nutrient enrichment can last several days after an iceberg passes through a stretch of ocean. Scientists estimate that much of the freezing Southern Ocean is enriched by the icebergs that float through its waters. A team of oceanographers and chemists in California has said that the "iceberg zone of influence" extends tens of kilometers away and lasts several weeks.

In the Arctic, icebergs create comparable hotspots of chemical and biological enrichment. Though the species supported by the iceberg might vary, the resulting web looks similar. Zooplankton including copepods and larvaceans graze on the phytoplankton that are fed by the iceberg and become prey for chaetognaths. Eventually, perhaps, seals and polar bears depend on icebergs not just for the life they attract but also as resting grounds while they skulk about the Arctic.

These iceberg ecosystems are connected to larger planetary ecosystems. By converting energy through photosynthesis, phytoplankton absorb carbon dioxide. When shrimp-like krill eat the phytoplankton, they eventually excrete the remains, causing the carbon to sink deep into the ocean. Icebergs, therefore, are indirectly responsible for drawing carbon away from the air and burying it in the ocean. In a 2013 study, scientists claimed that "sedimentation associated with the expanding number of free-drifting icebergs have the potential to increase the drawdown and sequestration of CO_2 in the Southern Ocean, thus impacting the global carbon cycle in possibly significant proportions." Or, as others have simply put it: Antarctic krill are climate heroes. A similar effect is hypothesized in the Arctic, since icebergs create conditions that allow consumers to "convert and package" primary producers like phytoplankton to "particulate organic carbon as fecal material." In 2016 researchers concluded that icebergs lead to millions of tons of carbon being trapped and buried into the ocean. Icebergs, in sum, play an important role in the oceanic carbon cycle and are slowing climate change.

Towing an iceberg out of the polar regions and to a new location changes the natural course of this mini ecosystem. If an iceberg is towed at a faster rate than it would normally travel and on a different course, how would that influence the community of organisms that normally thrive around the ice? How would it alter the transformation of carbon dioxide in the Arctic and Antarctica? Disturbing the smallest part of a web can have consequences for the whole structure. For the same reason, moving an iceberg outside of its natural path might disrupt bigger ocean creatures with troubling consequences. For example, it might relocate phytoplankton, which would eventually diminish the hunting grounds for orcas in the Antarctic, or relocate the hunting grounds for bowhead whales in the Arctic, which rely on plankton patches as they migrate to the Pacific. If enough invisible chemicals are moved along with icebergs, could the ripple effects be felt throughout the oceans until a disoriented, starving whale washes up on a beach far from the poles?

Dr. Mosley-Thompson cautions against this is hyperbolic thinking. Towing away one iceberg, after all, would not fundamentally or permanently alter the oceanic ecosystem. Other big-thinking scientists agree, including Dr. Waleed Abdalati, former chief scientist at NASA. Dr. Abdalati is now the director of the Cooperative Institute for Research in Environmental Sciences at the University of Colorado Boulder and studies how and why the earth's ice cover is changing. "My knee-jerk reaction," he tells me when I ask him about the environmental consequences of iceberg towing, "is that there will be little consequence." Dr. Claire L. Parkinson, a senior climate change scientist at NASA who has done fieldwork in both polar regions, had the same response. "The impact might not be great because of scale. There are thousands of icebergs in the Antarctic and the number removed might not make a difference." Dr. Parkinson has been with NASA since the 1970s and is the author of more than one hundred scientific articles about the Arctic and the Antarctic, with a special emphasis on polar sea ice and its connections to global climate systems. Dr. Parkinson—who became fascinated with Antarctica shortly after the time when women scientists were not allowed on American expeditions to the Southern Continent—also notes that some animals might actually be pleased to see the icebergs go. Penguins, for instance, find it harder to climb on icebergs than sea ice. Removing them might make it easier for the flightless birds to navigate where they would like to go. "Seals could benefit even more than penguins," she muses.

Dr. Abdalati and Dr. Parkinson become more circumspect when I invite them to imagine hundreds or thousands of icebergs being harvested. They admit the effects could be dramatic if this were the case. Dr. Parkinson further observes the presence of ships would increase the risk of oil spills and environmental disasters associated with human-made machines. Neither scientist imagines a beached whale halfway around the world, though. Yet, without exception, every scientist I talk with says that we need more information about iceberg harvesting.

Studies would also be useful to test the validity of some of the more outlandish claims iceberg entrepreneurs make about the potential benefits of their proposed endeavors. Several of the would-be harvesters allege, for instance, that collecting icebergs would help combat sea-level rise. Dr. Mosley-Thompson is quick to disprove this point. "Melting icebergs do not affect sea level as they are already floating and the density difference between ice and water also ensures there is no additional mass to the ocean when floating ice (i.e., icebergs) melts." She makes an analogy to a glass of water with ice in it that is filled to the brim. The ice melts but the glass does not overflow.

By removing giant icebergs from the ocean before they melt, some entrepreneurs say, they are mitigating the influx of meltwater pouring into the Arctic and the Antarctic. The harvesters make a related and slightly more grandiose contention about their ability to help keep ocean currents in balance. This relates to the Beaufort Gyre Current, which stores freshwater near the surface of the ocean in the Arctic. In the past two decades, this freshwater reservoir has increased by 40 percent. Normally, the current slowly releases the water through the Labrador Sea into the North Atlantic. If too much freshwater were suddenly to enter it might prevent the cold water from the Arctic from sinking deep into the ocean, thereby slowing the Atlantic Meridional Overturning Circulation, which carries warm water from the tropics northward. This, in turn, could influence the climate of the Northern Hemisphere. Removing icebergs and the freshwater they contain, the argument goes, would help maintain the delicate equilibrium upon which the existing climate system depends. These intrepid businesspeople are helping to ensure Western Europe remains pleasantly warm for its latitude.

For icebergs to help combat this freshening, a mind-bending number would have to be removed. The same can be said of the claim that harvesting icebergs would combat sea-level rise. "If that were the case," Dr. Parkinson tells me, "we should probably be concerned about the negative environmental costs as well." The scientists I interview are collectively suspicious of the purported benefits of iceberg harvesting but will not reject them outright. After all, these big thinkers have made careers understanding the interconnectedness of our planet and they know that icebergs, whether floating through the ocean on their own volition or behind tugboats, are proof.

◆

Towing an iceberg to people in need of freshwater will not just affect polar ecosystems and the ocean through which it is dragged—it will also influence the local environment where the ice is deposited by changing the temperature and salinity of the water, potentially altering the air mass, and introducing a large physical presence to a new location. Estimating the environmental impact of this invasion is near impossible since the resulting harm would be site specific. Still, damage to marine life would likely be unavoidable. Dr. Mosley-Thompson notes that the plant and animal species living off the coast of Cape Town or Fujairah would not be expecting an influx of cold water or freshwater. Benthic, bottom-dwelling organisms and pelagic communities, including fish, marine mammals, invertebrates, and sea turtles, could be affected. How each individual species responds further depends on the size of the iceberg and the species' own adaptability. Heiner Schwer of POLEWATER dismisses this concern, assuring me that "the problem about changing the water in the place it is going is very small." The effects, he insists, would be hyperlocal. "Within a few hundred meters of the iceberg, the water conditions would return to normal." Dr. Abdalati tends to agree. He points out that icebergs would be towed to ecosystems that are less fragile than the Arctic and Antarctic and that the introduction of freshwater in small doses could be inconsequential.

Besides the introduction of ice itself, the infrastructure required to harvest the freshwater that escapes the ancient block has the potential to damage the environment since the berg will need to be secured in place. There are currently three main methods under consideration: building a fixed platform, bringing weights into the ground, and suction buckets. Building a fixed underwater platform to hold the iceberg, like those supporting a majority of offshore wind farms, involves pile-driving irons into the seabed, which would upset the life it contains. The noisy hammering of construction, moreover, has the potential to disrupt marine species, such as whales. Construction of offshore wind farms and the underwater noise and the vibrations it causes has been blamed for the stranding and death of mink whales in the North Sea. Anchoring the iceberg to the seabed would create less immediate damage to the local environment, but would likely require lots of concrete. New research shows that making cement for concrete releases 8 percent of the total greenhouse gasses that go into the atmosphere annually—more than four times as much carbon pollution from the aviation industry. Even if an old submarine channel is used, like Nick Sloane hopes to employ off the coast of South Africa, the ice would still need to be anchored. The German company's suction buckets purportedly do little harm to the environment since they rely on the pressure difference generated between the inside of the bucket and the surrounding water to install the structure without the use of any mechanical force. Even that option will result in some operational noise, which could emanate through the water column, potentially injuring or even destroying sea life. But Schwarzer clarified that POLEWATER has invested in bathymetric studies and created a reversible system.

There could be effects outside of the water, too. Anchoring an enormous chunk of ice in warmer waters, whether in South Africa or the United Arab Emirates, could alter weather patterns, or so Alshehhi asserts. He imagines clouds forming from the iceberg that could transform the desert into an oasis. Dr. Timothy Cronin, an atmospheric scientist who has devoted his professional life to studying clouds, is skeptical of these claims. Dr. Cronin explains "Icebergs *could* make fog"—that is, a cloud at ground level—"if they were big and were placed in a warm and humid environment like the Persian Gulf." When moist air masses at different temperatures meet, the mixtures of air are supersaturated and form cloud droplets. The MIT professor notes that the resulting fog "would be cold compared to the surrounding air over the warm ocean, so it would sink and spread out over the surface—and it would not be likely to somehow rise high up into the atmosphere and trigger rain formation." Without prompting, Dr. Cronin is willing to stretch his imagination. "If the iceberg were big enough, though, then maybe the spreading out of cold air would be fast enough to force other air from over the warm ocean to rise upwards and form rain clouds." He concludes by restating his skepticism and his desire to see some sort of simulation or experiment as proof of concept.

If clouds did form thanks to relocated icebergs as Alshehhi hopes, it would not necessarily be good news. Previous efforts to create clouds have caused flash floods and mudslides that left several people dead. Understandably, many environmentalists caution against weather modification and cloud seeding in particular. Towed icebergs are meant to bring relief to the desert, not unleash disaster.

Beyond causing damage to the local environment, harvesting icebergs may contribute to the ever-worsening climate disaster by necessitating considerable energy expenditure to rearrange the planet. Dr. Mosley-Thompson points out the biggest environmental issue with iceberg towing might be the carbon footprint such a hubristic act would leave behind. Greenhouse gas emissions are difficult to calculate, given the hypothetical nature of towing. Much would depend on the size of the vessel, the type of fuel it uses and the efficiency of its engine, the power it exerts at sea and how long it sails, as well as the size of the iceberg.

Regardless, it does not take much imagination to realize that dragging an iceberg from an Antarctic island to Fujairah or Cape Town or from Nuuk to the Canary Islands could generate tons of CO_2. A single cruise ship can produce more carbon dioxide than 12,000 cars. According to the *European Maritime Transport Environment Report 2021*, a cruise ship emits 10 tons of black carbon annually, and a container ship generates 3.5 tons. Even a small vehicle carrier produces 2.1 tons each year on average. Climate scientists estimate that 1 ton of CO_2 emitted into the atmosphere destroys three square meters of the Arctic summer sea ice. Added CO_2 would also harm places like Cape Town, where climate change is exacerbating the freshwater crisis. Paradoxically, iceberg harvesting could contribute to the underlying problem for which it is meant to be a solution.

The people who are already selling iceberg water are cognizant of the environmental costs of their businesses, and most have taken proactive steps to counter the harm, or at least the bad press it could inspire. Svalbarði, for instance, offers financial assistance to update water infrastructure in Uganda and Malawi and supports a western China wind farm; as a result, it has been certified carbon-neutral. Actually, Svalbarði is carbon-negative since it devotes a significant portion of its revenue to supporting CO_2-reducing projects. Jamal Qureshi claims that his company, on net, removes CO_2 from the atmosphere.

Qureshi is offended when I suggest he is engaging in "greenwashing," or marketing that makes misleading claims about a company's sustainability to convince consumers of its environmental friendliness. For what it is worth, when I meet with him in Norway, it is evident that Qureshi loves the Arctic and genuinely wants to help protect the ecosystem. His eyes sparkle as he discusses the beauty of Svalbard, and he reminds me that he relocated his family to live year-round in the polar outpost. The latest tagline for Svalbarði is "Taste the Arctic to Save the Arctic." Qureshi is giving up much of his own profit to protect the environment the best way he knows how. Still, one has to wonder about the carbon footprint left from a heavy glass bottle that is shipped all the way from Longyearbyen to Los Angeles, Dubai, or Shanghai.

Carbon offsets might not be the answer to making iceberg towing green. Buying the credits has been plagued by problems when it comes to reducing emissions that would not have happened otherwise. In some cases, companies buy carbon credits for old projects that produce far more CO_2 than newly built green technology would. The money, in other words, could be better invested to help combat climate change, though it would not come with a handy carbon-neutral certificate. More problematically, it is possible for companies to purchase carbon credits that are not "additional"—that is, they would not have occurred in the absence of a market for offset credits, as the Stockholm Environment

Institute explains. If the reduction would have happened or did, in the case of an existing wind farm, say, then offsets purchased are not additional. And if greenhouse gas reductions are not additional, a company that buys credits instead of curbing its own emissions will actually worsen climate change.

It may be impossible to move icebergs around the planet without creating some greenhouse gas. Prince Mohamed Al-Faisal's original idea in the 1970s, to attach nuclear plants, might actually be greener than using existing fuel-powered tugboats and purchasing offsets. Of course, we might not want floating, mobile nuclear power plants on the stormy seas, either. There is enough traffic on the ocean without potentially introducing radioactive waste to the waves.

Moving at a turtle's pace, the tugboats dragging icebergs through the ocean may not seem like terrible dangers to others at sea. However, the boats will have limited ability to steer and be unable to make quick changes to their trajectories with a fragile iceberg fastened to a rope a mile behind them. For safety reasons, the teams at the Southern Ice Project and POLEWATER also acknowledge that they may have to disconnect from an iceberg they have harvested. Or the high waves and breakneck winds in the Southern Ocean could cause the iceberg to roll loose. Or the iceberg might disintegrate into multiple growlers and bergy bits like a cluster bomb, threatening nearby ships or oil platforms. We know from centuries of experience that ships sink and people die when they unexpectedly encounter icebergs. Even when they are on the lookout for the frozen menaces, captains cannot always avoid them. Even when they can, we know from the International Ice Patrol that not all ships are happy to navigate around areas rendered unsafe by the ice.

The likelihood of a collision in such a circumstance is not so far-fetched. On maps, we often see the ocean as one undifferentiated expanse of blue. In practice, it looks a lot more like a road atlas. Currents, weather patterns, prevailing winds, underwater topography, and safety concerns have created established optimal routes that function like highways with multiple lanes that move in specific directions. And these shipping lanes are crowded. The Cape of Good Hope, for instance, maintains the same geostrategic position it held in the fifteenth century, lying on the sea route that connects the Asian, European, and American markets. Around twelve thousand cargo vessels use South African ports every year. Introducing a dangerous iceberg-dragging vessel will only further compromise the safety of the ships sailing along the Benguela Current.

Some of the environmental issues may overlap with such logistical problems. Lugging icebergs into shipping lanes, for example, may attract more vulnerable animals to the busy oceanic highways and put those animals at risk of collision. The International Maritime Organization, which governs shipping worldwide, recently altered the permissible routes off the coast of California to balance the safe and efficient flow of commerce with animal safety since the previous lanes sliced through an area in which endangered whales fed. As sea traffic increased, so did accidents with the marine mammals. Adding icebergs into the mix might only further complicate matters.

◆

The North and South Poles are among the most rapidly changing parts of our world. Sea ice is vanishing, glaciers are melting, and permafrost is thawing as human-made emissions drive global temperatures up. In this context, it is difficult to claim that removing icebergs is more or less natural than what is currently happening in the Arctic and the Antarctic. Perhaps we might as well harvest icebergs, even if it contributes to climate change and potentially damages ecosystems in the polar regions and beyond. The risk may be worth taking because it is hard to know the magnitude of the environmental consequences of iceberg towing.

Dr. Abdalati helps me understand why more people have not studied this issue. "There is a little bit of the fantastical dimension," he explains. "And the scale issue. Plus, who is going to fund it? It is very high-hanging fruit. The effort to understand the environmental consequences are not commensurate with the rewards for figuring it out." And most scientists cannot conduct research without funding. Plus, as Dr. Abdalati observes, the people who are financially backing iceberg harvesting are not necessarily going to pay for the research because they may not want the results. Or, like Dr. Orheim says, "We can figure out the environmental consequences once we have actually accomplished the feat." It will be easier to secure funding once we show it is possible to move icebergs to Cape Town, Fujairah, or the Moluccas.

Harvesting icebergs might be easy to greenlight because the exotic marvels seem so far away. Deep ecologists like Val Plumwood identify "remoteness" as one of the problems of global climate governance. Humans, particularly socially privileged humans, experience high-level dissociation between the costs and benefits of their ecological consumption. Remoteness separates us from the immediate outcomes of our commodification of the natural world spatially, temporally, and consequentially. In terms of iceberg harvesting, the environmental harms to the Arctic and the Antarctic are far away from those of us in places like New York, Berlin, and Hong Kong. The damage iceberg towing would cause might also be temporally remote insofar as the ecological effects will be felt in the future and not by us now. And, for those of us who can better afford to comfortably adjust to a warming and more volatile planet, the consequences of pursuing an activity that further contributes to climate change might be consequentially remote because the worst effects will likely fall on other peoples who do not have the resources to turn up their air conditioners or relocate when desired. Remoteness thus makes it easier to pursue the commodification of nature and overlook the environmental consequences.

Icebergs are not remote for everyone, though. Nor are they merely frozen components in complex polar ecosystems. For peoples living in the Arctic, icebergs are omnipresent. They do not have to fantasize like Melville's Ishmael about icebergs, and they do not need to gawk at promotional videos to imagine what it might be like to see an iceberg up close. Harvesting icebergs would take a part of the Arctic peoples' home and relocate it thousands of miles away. In so doing, we would not just be mixing up the world ecologically but also culturally.

6

Situating Icebergs

In Greenland, it is hard to imagine any problem with taking an iceberg and hauling it off elsewhere to help combat water insecurity. There are so many pieces of ice floating around the island, removing one the size of the Empire State Building would go unnoticed. The country's third largest city, Ilulissat, even means "icebergs" in Greenlandic. Home to around 4,500 people and 3,500 sled dogs, the town sits 220 miles north of the Arctic Circle. To get here, visitors must travel by plane, boat, snowmobile, or sledge since no roads connect settlements in Greenland. The houses in Ilulissat are painted bright blues, oranges, and yellows and are seemingly arranged at random on the pale gray gneiss and granite that dominate the landscape. Sled dogs howl throughout town. And everywhere I look there is an awe-inspiring view of snow-covered mountains or ice-filled water. This is thanks to Sermeq Kujalleq, the glacier that lies some thirty miles away deep inside the Ilulissat Icefjord.

Tourists are greeted by a foreboding sign: EXTREME DANGER! DO NOT WALK ON THE BEACH. DEATH AND SERIOUS INJURY MIGHT OCCUR. RISK OF SUDDEN TSUNAMI WAVES, CAUSED BY CALVING ICEBERGS. Sermeq Kujalleq hurls icebergs into the sea at a stupefying rate: more than eleven cubic miles of ice annually.

If melted, it could provide enough drinking water for everyone in the United States for a year. Some of these bergs are as tall and blocky as skyscrapers. Others look like crystal palaces and cozy cottages. In between, bergy bits and growlers churn like packing peanuts. It takes my eye a minute to perceive that these icy masses are slowly moving.

People have long interacted with the icebergs in Ilulissat. Abundant fish in the fjord have attracted hunters to the small meadow that hugs the rugged coast since the Stone Age. First the Saqqaq people lived at the mouth of the fjord at Sermermiut, followed by the Dorset and Thule peoples. European whalers arrived in the sixteenth century and in 1741 established a nearby trading post in Disko Bay that would become the colony Jakobshavn. Around 1850, the last people moved from Sermermiut to the colony, which was renamed Ilulissat in 1979 when Greenland secured home rule, established its own parliament, and gained sovereignty over its environmental policies. Now, Greenland plans to further bolster its financial independence from Denmark by selling water to the highest bidder, which may include icebergs. Given the abundance of ice, the scheme should work well in theory. Yet ambivalence reigns.

For many people in Greenland, icebergs are more than just giant abiotic components in a complex polar ecosystem. That does not mean, however, that they view the ice as exotic alien apparitions, like many tourists in Canada do. For the folks living in Ilulissat, icebergs are a beautiful if quotidian backdrop for daily life.

Anthropologists are trained to ask questions about how meanings are constructed and to reflect on the situatedness of life. They consider how linguistic, sociohistorical, cultural, economic, geographic, political, and material dimensions influence how they and others see the world. Because of our situatedness, I likely have different ideas about water than do people in Greenland or

South Africa. The same concept applies to objects, too. Identical water—H_2O—from a chemical perspective, takes on a different meaning in a water bottle versus a swimming pool. Similarly, a body of water in my home state of Minnesota, the Land of 10,000 Lakes, has a different connotation than it would in the Arabian Desert. Accordingly, scholars of water often ask: "How is water situated?" They mean both geographically and culturally. They consider the physical topography of water, as well as its historical context, the existing social and power relations around it, and the philosophical frameworks underpinning perspectives on the water to determine its situatedness.

For much of this book, I've looked at icebergs from a Western perspective—after all, it is my own point of view. This outlook, for better or worse, also largely informs the way many international organizations, like the United Nations, function and is thus worth well understanding. Only considering icebergs from this vantage point, however, would result in a myopic look at the important resource. In Greenland, I attempt to reflect on the situatedness of icebergs and learn about them from an alternative standpoint. It helps me question whether icebergs are part of the global commons that might be used to help alleviate water insecurity in places like Cape Town or Fujairah or Los Angeles. Although all life, human or otherwise, depends on water, not everyone has the same relationship with the resource. This is certainly the case with icebergs. The frozen freshwater is cultural and the people in Greenland have taken care of it for generations. Even if we decide to take the environmental risk of harvesting and towing icebergs, we ought to also consider the human costs and who should get to benefit from this water. I get help thinking about these issues in Ilulissat.

◆

Kistaaraq Abelsen laughs when I ask if I can ask her a stupid question. We're sitting in a café near the Ilulissat Icefjord, which recently opened as part of a new visitor center at the UNESCO World Heritage Site. Practically the whole town attended a *kaffemik* to mark the opening. Traditionally, Greenlanders open their homes to guests for an entire day to celebrate special occasions such as birthdays and graduations. Family, friends, and acquaintances stop by throughout the day to enjoy cakes and delicacies made by the host. In my trips to Greenland, I've been invited to several *kaffemiks*, including once after briefly chatting with a man who saw me sliding on my backside down a mountain after a hike got too steep. As he continued up the snow-covered rocks, the stranger turned around and said it was his son's confirmation and that I should visit. It is possible he felt sorry for me, but I think the invitation stemmed from genuine cordiality since Greenland tends to be an exceptionally hospitable place. The Icefjord Center certainly wanted the whole town to feel welcome in the new space. It seems to have worked; there are more Greenlanders than tourists in the café when I visit. Following local custom, our shoes are sitting by the front door, which makes chatting with Abelsen feel all the more intimate.

"Sure," she says warmly. Kistaaraq Abelsen is a short woman with black hair, a glowing complexion, and a big smile. I apologize for my imperfect Danish and she needlessly excuses her English. We proceed in a mix of the two languages. I've learned to approach my question from several different angles and languages to try to gain traction. A friend who teaches at the University of Greenland warned me that many people would not understand my question. "It just wouldn't occur to most people," he explained. "Or they might think it is a joke." Selling icebergs in Greenland is a bit like peddling dandelions in the United States. Most people would not think of them as a commodity. I've already struck out several times in Ilulissat trying to ask people about icebergs, so I am hesitant yet another Greenlander will think I'm not in my right mind.

"What do you think of selling icebergs?"

Silence.

I start to repeat myself in Danish and Abelsen cuts me off with a friendly chortle. "I understood you!" she exclaims. "It is just a tricky question."

I'm relieved. Abelsen grins and adjusts her glasses. "Icebergs aren't just icebergs," she says. "They are part of Greenland, part of our culture."

This is especially evident in Ilulissat, where an entire tourism industry has developed around Sermeq Kujalleq. Each year around fifty thousand people travel from all over the world to marvel at the ice. They can take hikes, boat trips, and air safaris to see the giant bergs. One company offers visitors the chance to kayak and paddleboard among the ice. The new Icefjord Center is meant to attract still more tourists to the region and help them understand the breathtaking landscape.

"It is deeper than just the scenery, though" Abelsen explains about the connection to icebergs. "Icebergs are part of our mythology and identity." Abelsen is a math teacher originally from Sisimiut, a town 170 miles away. She's been teaching in Ilulissat for a year and is about to move to Southern Greenland for another teaching position. I suggest that especially for people from Ilulissat icebergs must be special. "They are important to all of us!" It is the only time Abelsen is at all sharp.

◆

To begin to understand the importance of icebergs in Greenland, one need look no further than the flag. Adopted in 1989 after Greenland achieved home rule from Denmark, *Erfalasorput*, or "our flag," as it is known in Greenlandic, is split into two equal horizontal bands. The top is white and bottom is red. A large circle stands just left of center. The top is red against the white background, symbolizing the rising sun over the permanent ice, and the bottom semicircle is white, representing an iceberg in the ocean. It took more than a decade and hundreds of proposals for this flag to be selected. Most other options included a Nordic cross, familiar on the flags of Denmark and its Scandinavian neighbors. The newly formed Greenlandic parliament opted for the iceberg, instead. It is truer to Greenland.

For as long as people have lived on the world's largest island, they've lived among icebergs. As a result, icebergs are ubiquitous in Greenlandic culture. This is perhaps best recorded in Greenlanders' rich mythology. Until Danes came and started writing down stories—and freezing them, for better or worse—Greenland was an oral culture, passing stories down and modulating them, generation to generation. Expectedly, icebergs are strewn throughout the myths. They are a place for talking ravens and polar bears to rest. Hunters climb icebergs as a test of their skill, and runaway children take refuge on them. Icebergs create exciting adventures, like in one Greenlandic tale of the famously adventurous Kivioq, who must quickly paddle through the opening created by two jostling icebergs. In many myths, icebergs are a prop, like in the story of Qasiagssaq, the great liar, who, to gain sympathy from his wife, fills his kayak with pieces of iceberg and even stuffs his clothes with ice shards to make it seem like an iceberg calved on top of him. In others, powerful figures summon icebergs for good, like bringing seals, or conjure them for evil, to crush houses and kill people.

In several myths, icebergs are a vital water source. In the tale of Tiggak, the protagonist is an *angakkoq*, or shaman, who does not fit in with his wife's family. One day, Tiggak gets stranded on the hunt with his brothers-in-law. When they grow desperate with thirst on the open ocean and are ready to die, Tiggak sings a song to summon an iceberg. Sure enough, a berg appears. Tiggak sings another song, and a spring of delicious freshwater bubbles forth from the iceberg. The men are saved, and after a series of further adventures return home.

In the nineteenth century, Hinrich Johannes Rink, who first recorded many of the myths, had one character say it straightforwardly: icebergs are "the fountain designed for those perishing at sea." Indeed, during the summer, kayakers on the salty ocean would traditionally rely on icebergs for freshwater. In winter, too, Greenlanders used icebergs as a water source. When the inland ice stopped melting and froze fast, they could find a lodged iceberg and cut away a chunk to bring home, where they would let it melt in a water pot.

Icebergs turn up across the arts in Greenland. From nineteenth-century woodcuts by Aron from Kangeq, to Hans Lynge's twentieth-century impressionist oil paintings, to contemporary digital photo collages by artists like Ivinguak Stork Høegh, icebergs glisten, often as a shorthand for Greenland itself. Folk singers like Rasmus Lyberth croon about icebergs and even the Greenlandic rock band Nanook features loving shots of icebergs in their music videos. It is no surprise, then, that icebergs are found in *Nuummioq*, the first feature-length film produced entirely in Greenland. After its premiere at the Sundance Film Festival in 2010, Reuters called the film, about a man named Malik coming to terms with his cancer diagnosis, a "stunning triumph." For my taste, it is a little depressing, but I was charmed by a side story involving the protagonist's scheming cousin Michael, who has a plan to sell icebergs. "Here it's like robbing a bank," he says to Malik, "except it isn't illegal." The family laughs at Michael, who persists: "People will love it. We just have to sell it the right way. If we export it, it'll be luxury goods. We'll be rich." Michael has even dreamed up a commercial to sell his product: a rich guy in Bermuda drinking whiskey in need of ice cubes from icebergs. Everyone laughs. Selling icebergs is a punchline that is meant to show that Michael is a good-for-nothing and that Westerners are idiots. With so many icebergs in Greenland, no one here would buy them.

◆

"Are people idiots for buying icebergs?" I provoke. Abelsen doesn't answer. She is not the type to characterize anyone that way—not in front of me, at least. Instead, she answers my question by telling a story, explaining that last week it was her sister's birthday. They threw a big party and needed ice. So Abelsen went down to the beach, scooped up a growler, and brought it home. After letting it melt for a minute, she broke it up with a chisel, and voilà, fresh ice for the party guests. "So *you* would be an idiot for buying ice-bergs," I interpret. She laughs and nods. Almost everyone I talk to in Greenland uses icebergs this way. They collect them in boats and bring them home and plop them in the freezer for later use.

Abelsen initially dismisses the idea of iceberg harvesting as a childlike fantasy, like the plan in *Nuummioq.* "You don't have icebergs like we do, though" she says, sipping her hot coffee and looking toward the bergs slowly flowing through the fjord. She understands why rich people in Los Angeles or Shanghai would pay for them, even if she says paying $100 for a bottle is absurd. "They are special," she sighs, "even to me." As Abelsen explains it, icebergs are always changing, so they are always a compelling sight. When they twist in the water, they catch the light at new angles, twinkling like diamonds, and constantly recreate the landscape as they capsize and break apart. "They still take my breath away," she says. The sentiment is shared by the other residents of Ilulissat I interview, including the twenty-year-olds I meet at Bar Naleraq and the old fishermen who work catching halibut and prawn for the Royal Greenland Seafood Company. Only one young woman I spoke to had something negative to say. "I'm so sick of icebergs," she moaned. "You can have them."

Although Abelsen claims to be unbothered by the idea of a chump buying icebergs, I am curious whether her permissive attitude applies to a hypothetical merchant, too. After all, if somebody is buying, someone is selling. I press: "What if an individual started collecting and selling hundreds of icebergs in Ilulissat?" Abelsen's nose wrinkles. "But they belong to all of us," she murmurs. She begins to identify several potential problems. She notes that Greenlanders rely on icebergs to attract tourists. Too many boats collecting bergs offshore could spoil the view. Then there is the matter of the environment. It can't be good to drag icebergs all over the planet, Abelsen notes. Something about one person getting rich from a shared resource also doesn't seem to sit right. I want to understand how the deep cultural connection to icebergs resonates with the potential commoditization of the resource. If icebergs are a special part of Greenlandic culture and something that people treasure, how does Abelsen feel about people selling them?

"We could do it like with the seals," she answers. Seal hunting has a long tradition in Greenland and remains, in the words of an official government report, "a vital component of everyday life and culture." Seal meat can purchased in almost every grocery store, its skin is used as part of the national costume and for hunting equipment, and the fur is made into clothes and accessories, like the many stylish fanny packs I spotted on my most recent trip to Greenland. In 2009 the European Union banned the import of sealskin products on the basis of animal welfare. This created huge problems in Greenland, since many people rely on seal hunting and some communities are fully dependent on trade in sealskin. Six years later, the European Union made an exception for sealskins from hunts conducted by Inuit and other Indigenous communities. Now, Great Greenland, a corporation owned by the government, sells beautiful seal products to the European Union. The skins are caught by Inuit hunters up and down the Greenlandic coast according to Greenlandic custom. The hunt is therefore touted as both ethical and sustainable, since there is never an overproduction and the seal population is protected. This model allows Greenlanders to continue to engage in their traditional practices while participating in global commerce. It also doesn't make the seal any less special in Greenland. People can partake and profit while still safeguarding their customary practices.

When our discussion returns to icebergs, Abelsen states that "people just shouldn't take too many." I agree and ask an unfair question: "How many is too many?" The math teacher admits she doesn't know, and she doesn't seem too interested in figuring out an answer. But her proclamation betrays her belief that icebergs could be improperly exploited with regard to the environment and the people of Greenland. When I ask if she trusts her government to make that determination, Abelsen crosses her stockinged feet under the table and gives a coy look. She doesn't want to answer, and I don't make her.

She adds a caveat of her own. "Before Americans use our icebergs," she says, "they should come to Greenland and see who we are and how we live. They should try to understand us." She pleads, in other words, for greater cultural understanding. I think back to 2019, when President Donald Trump joked that he should buy Greenland, treating the country and its people as a resource to be exploited. I mention Trump and Abelsen puffs her cheeks and rolls her eyes. I laugh and we change subjects.

◆

In Nuuk, home to approximately fifteen thousand people, I learn the official government position on icebergs. Anja Sørenson, a special adviser in Greenland's Ministry of Industry, Energy and Research, is leading the effort to sell more water. She sits inside the Nuuk Center office tower, which houses the country's government and a shopping mall. The ten-story glass-and-steel structure is the tallest building in the country and looms over the capital city. Its matte white façade is meant to symbolize snow and icebergs. The building unintentionally represents Greenland's political predicament, too. The modernist design seems to promise a bright future, yet it was constructed by a Danish firm using five thousand tons of concrete and one thousand tons of steel that had to be imported to Greenland by ship. The survival of Greenland without Danish subsidies is uncertain unless the country can exploit its own resources. Luckily, Sørenson has a plan. With her blond hair pulled back into a tight ponytail and attired in a stylish dark suit, the Danish-trained lawyer projects matter-of-fact confidence. "Why not use icebergs?" she asks when I first meet with her in 2019. "They'll just become salt water."

Using icebergs as a water source is nothing new in Greenland. Just as locals in Ilulissat and throughout Greenland use icebergs for personal consumption, the government utilizes icebergs in their public water works. In Qaanaaq, a settlement 867 miles from the North Pole where some 650 people live, it is so cold that for most of the year, rivers are frozen fast and little freshwater escapes the inland ice. Tanks are filled in the summer, but don't hold enough water to get inhabitants through the winter. Consequently, the town sends dump trucks out onto the frozen sea to gather icebergs. Back in town, the trucks deposit the ice into a large smelter, filter it through sand, treat it with ultraviolet rays, and add chlorine. Then it joins the rest of the water system and flows from the taps inside the colorful homes that stand on the permafrost. Icebergs, in sum, have long been understood by government officials as an exploitable resource.

In the time I've spent in Greenland, I have learned it is one thing when Greenland uses its own resources for its own people and quite another when outsiders get involved. Greenland has a troubling history in this regard. Modern Danish colonization of Greenland began in 1721, when private merchants traded European goods for Greenlandic commodities, including whale blubber, sealskins, and narwhal teeth. By 1776 the Danish Crown assumed control and granted its Royal Greenland Trading Department a monopoly on trade to and from Greenland. To maintain its exclusive stranglehold on the island's resources, Denmark claimed to be safeguarding traditional Greenlandic culture from corrupting outside influences. Under the same justification, Denmark compelled Greenlanders to remain in occupations like hunting and fishing. The state also pursued a so-called civilizing mission, including forced conversions to Christianity, the suppression of language and traditions, and the destruction of existing communal structures. According to the prevailing Danish rationale, Greenlanders were naïve and child-like, in need of protection and incapable of governing themselves. In reality, the Danish government wanted to ensure that it could exploit the vast Arctic resources however it liked.

These policies persisted until World War II, after which the United States, eager for a piece of Greenland's natural riches and with an eye on its geopolitical advantages, offered to purchase Greenland for $100 million in gold. Denmark rejected the offer and elevated Greenland to an official county, or *amt*, with representation in Parliament. The subsequent "modernization" or "Danization" of the island, through the 1950s and 1960s, caused more colonial damage. Danish efforts to industrialize cod fishing required concentrated populations rather than dispersed hunters, resulting in the forced relocation of Greenlanders, rapid urbanization, and a restructured labor market. Danes were imported to the island to run businesses; they grew from 5.25 percent of the population in 1950 to nearly 20 percent by 1975. New laws dictated that civil servants born in Greenland could be paid only 85 percent of the salary of those born in Denmark. Families were split apart; in some cases, children were sent to Denmark for schooling. Increased suicide rates, alcohol abuse, and violence followed.

For the past four decades, Greenlanders have been working to regain their autonomy. In 1979 Greenland won home rule, and in 2008 Greenland's citizens voted for self-government in a public referendum. Today, Greenland controls its legal system, police, internal affairs, and natural resources. Greenlandic has replaced Danish as the official language. Under international law, Greenlanders are recognized as a separate people. In a recent poll conducted by the Universities of Greenland and Copenhagen, more than two-thirds of Greenlanders expressed a desire for their country to become independent, most of them within twenty years. Understandably, Greenlanders are wary when outsiders desire their country's rich resources, even when the sale of those resources seems to be a path to independence.

In the most recent parliamentary elections in 2021, the left-wing environmentalist party, Inuit Ataqatigiit, won 37 percent of the vote to become the most powerful party in the country. They won after campaigning against mining projects backed by China and Australia—and supported by the incumbent center-left pro-mine Siumut Party—that sought to exploit rare earth elements on the island. Despite the promise of wealth and new jobs and increased independence from Denmark, Greenlanders rejected the projects out of concerns for the environment, their cultural heritage, and traditional livelihoods. For now, the mining projects have been halted. The election makes clear that exploitation of the island's natural resources will have to be on Greenlanders' terms.

When I check in with the new government about plans to sell icebergs, the mood has shifted a bit. This time, I talk with Marlene Kongsted, the new head of the Ministry for Business and Trade. She is initially more reluctant than Anja Sørenson to talk to me about icebergs and begins by telling me that government has different priorities now. Through an email exchange in Danish, Kongsted confirms that "there is still a focus on developing the area" of water sales and "the *Naalakkersuisut* is working on a new strategy in this area." Kongsted does not directly reply to my question about the new government's stance on how its environmental priorities might influence the sale of icebergs. Instead, she informs me that she can only answer once "we have a new strategy." Only time will tell what that might be, and how these traditional pieces of Greenland might be commoditized for the global market.

◆

Indigenous peoples across the globe have been dispossessed of water by state systems of water governance that prioritized and continue to prioritize non-Indigenous settlers as well as by private companies that accumulate water rights and use the resource for profit with indifference to who benefits. In the United States, Indigenous groups were forcibly removed from land coveted by settlers and forced to live on less desirable reservations, which often had limited access to water. In many cases the government then failed to invest in infrastructure to bring water to these places, like the Pine Ridge Reservation in the Badlands of South Dakota. To this day, some people on the reservation rely on delivered water or travel to a community spigot to fill up buckets to take home. Nationwide, Native Americans are nineteen times more likely than white Americans to lack complete plumbing in their homes. In other instances, the US government built dams and diverted water away from Indigenous peoples and completely left them out of water policy. For example, when seven states signed the Colorado River Compact in 1922 outlining how to "equitably" divide the river, Native Americans were excluded. So while Colorado was allotted some 51 percent of the Upper Basin and Arizona 37 percent of the Lower Basin, it was never settled how much water tribes could take from the river—a river on which they had relied for centuries. The same pattern can be observed around the world. Before the colonization of Australia, for instance, Aboriginal peoples managed and had access to 100 percent of the water in the all-important Murray-Darling basin in southeastern Australia. Today, some forty-four First Nations groups collectively share 0.12 percent of the basin's market value. In Bolivia, the privatization and commercialization of water sources has created dire conditions for Indigenous people. In Cochabamba, for instance, corporate control of water led to prices that the poorest residents could not afford and even charges to individuals, who sought to collect rainwater or draw water from their own wells. This injustice eventually led to the so-called Cochabamba Water War, which led in 2000 to the reversal of the privatization.

Among the chorus of activists and scholars calling for water justice, a handful are busy advocating for Indigenous water justice in particular. This entails acknowledging the right of Indigenous peoples to control, develop, own, and use water that is essential to their life and culture. These principles have been enshrined in international legal instruments, like the Declaration on the Rights of Indigenous Peoples, passed by the United Nations in 2007. Among other articles, the declaration acknowledges the right of Indigenous peoples to waters within their traditional territories and to create strategies of development and use for their lands and resources as part of their right to self-determination. Indigenous water justice calls for nation-states to delineate in domestic laws and policies Indigenous peoples' sovereign water rights and protect the right of Indigenous peoples to practice and revitalize water-related traditions. The principles also demand that domestic waters laws do not just grant a right to water but also ensure that there is adequate infrastructure to make those rights meaningful. Further, for advocates of Indigenous water justice, these rights include alienability and the ability to engage in water marketing. As part of the right to sovereignty, it is up to Indigenous peoples to resolve the tension, if any, that exists between the sacredness of water and the desire to participate in the growing water market.

Apart from recognizing Indigenous peoples' sovereignty and right to water, several international declarations acknowledge that Indigenous peoples have long been and continue to be caretakers of water. In 2003 Indigenous peoples from all parts of the earth signed the Kyoto Water Declaration. It states:

We were placed in a sacred manner on this earth, each in our own sacred and traditional lands and territories to care for all of creation and to care for water. We recognise honor and respect water as sacred and sustains all life. Our traditional knowledge, laws and ways of life teach us to be responsible in caring for this sacred gift that connects all life. Our relationship with our lands, territories and water is the fundamental physical cultural and spiritual basis for our existence. This relationship to our Mother Earth requires us to conserve our freshwaters and oceans for the survival of present and future generations. We assert our role as caretakers with rights and responsibilities to defend and ensure the protection, availability and purity of water.

In Greenland, this caretaker role is evident. I don't mean to imply that Greenlanders are like that terrible 1970s television advertisement featuring Iron Eyes Cody, a non-Indigenous actor dressed in a Native American costume who cries when he sees a polluted landscape. The romanticized idea of Indigenous peoples as pure and inactive agents within their environments—and thus natural environmentalists—is the product of settler fantasies that were used to justify colonization. And while many Indigenous groups do have a connection with the land that is distinct from a relationship based on imperialism and capitalism, the trope of the "ecological Indian" harmfully flattens the rich diversity of Indigenous life. Like the young woman I met in Ilulissat who couldn't care less about icebergs, far from everyone in Greenland is a climate activist. Yet, I observe that many Greenlandic politicians are cautious about embracing projects that would harm the environment and that the people elect them for that reason. I am also struck by how clean the landscape appears. The air is fresh, the ground is uncontaminated, and the water is pure. There is no wide-scale pollution or environmental devastation. For centuries, Greenlanders have protected the ice sheet.

Unlike for many Indigenous groups in the United States, Canada, and Australia who are still fighting to have their right to water recognized, the Inuit in Greenland make up the majority of the population and, after centuries of colonial wrongs, have growing political power to assert control over their natural resources, including water. Greenlanders, in other words, are poised to realize Indigenous water justice in a manner that is only a dream elsewhere. As a matter of Indigenous water justice, they should and can do whatever they would like with their water. Maybe that will include selling icebergs.

Leaving Greenland, I board a small plane that hops along the country's western coast from settlement to settlement, quickly letting passengers on and off. One Greenlander describes the experience to me as being more akin to a bus trip than an airplane ride. I step off the turboprop to stretch my legs in Aasiaat, a settlement between Nuuk and Ilulissat with a large seafood factory. Just feet away, a giant iceberg has been grounded on the small stretch of rocks that separate the short runway from the sea. The dozen passengers that file off the plane with me don't seem to notice it. Nor do the few people waiting at the picnic tables outside the airport building. Back on the plane, I realize I am the only person looking out the windows at icebergs, too. This view is apparently unremarkable.

Looking back, it is laughable how excited I was to first see icebergs in Newfoundland. My perspective was inevitably influenced by my cultural background, which reveres icebergs as mystical and exotic Arctic emissaries of a past pure world. This perception of the frozen masses is informed by geography, too. Only four hundred to eight hundred icebergs pass through Iceberg Alley every year on their migration to the wider Atlantic Ocean, and most cannot be observed from the coast. The few that sparkle within eyesight consequently seem like rare gems. It is why tourists flock to St. John's to go on sightseeing trips and why they buy iceberg beer, vodka, and water.

Gazing out the window of the small plane, it is clear how ubiquitous icebergs actually are. Embracing the outlook of my fellow passengers might make it easier to support the effort to drag an iceberg to Cape Town, Los Angeles, or Dubai. If there are so many around, why not use a few? Especially if these are not ephemeral jewels meant only for a select few, it is easy to imagine investing in transporting the resource to those who need it.

Yet, learning from Greenlanders also compels us to consider the situatedness of this resource, including the long colonial history of the island. I believe we cannot not just come and take icebergs to use however we would like. Greenlanders have a special relationship with the ice and it should be up to Greenlanders, as the custodians of this resource, to decide what to do with icebergs that come from the ice sheet here. It is thanks to Greenlanders that the icebergs that glide by Ed Kean in Iceberg Alley are pure, clean, and valuable.

Still, the icebergs I spy from the air are on their way to Canada. Greenland might be the birth site of these icebergs, but these wild beasts migrate. And icebergs are situated differently elsewhere. How do the claims of downstream users like Ed Kean square with demands for Indigenous water justice? Should the *Naalakkersuisut* be able to capture all of its icebergs and sell them to the highest bidder, even if it means that people in countries suffering from water scarcity would be excluded from the resource? What about icebergs in Norway and Russia and Antarctica? Although I came to Greenland hoping to find answers, I leave with still more questions.

7

The Law of Icebergs

For centuries, much of the Western world has operated on the principle of finders keepers, meaning that whoever finds something unowned or abandoned can claim it for themselves. The great Roman jurist Gaius wrote in the second century, "What presently belongs to no one becomes by natural reason the property of the first taker." As examples, he names wild beasts, birds, and fish, which were codified in the sixth-century *Institutes of Justinian*, and elaborates that anyone, Roman citizen or noncitizen, can acquire ownership by assuming control over such an unowned thing—what lawyers still today call by its Latin name, res nullius, meaning "nobody's thing." This ancient Roman law morphed over time into the common adage with the addition of "losers weepers."

One of the most famous losers in US legal history is Lodowick Post, who was out with his hounds hunting on an uninhabited beach in New York in 1802. Post was in hot pursuit of a fox and just about to seize the creature when another hunter, Jesse Pierson, killed the prize and took it home. Depending on whose story you believe, Pierson did or did not know Post was chasing the fox. Post sued, claiming the animal belonged to him. The lower court ruled for Post, and Pierson appealed, arguing that the plaintiff had no right to the fox merely by virtue of pursuing it. Nearly every law student in the United States reads this case because it succinctly demonstrates how things become property. In so doing, the case reveals property as a construct—an idea that we create and assign meaning. The majority in *Pierson v. Post* ultimately ruled for Pierson. Quoting the *Institutes of Justinian*, among other legal sources, the judges declared that a person does not gain possession of a fox by chasing it. Such a thing only becomes property once it is fully captured or killed. Finders keepers, losers weepers.

To the ancient Roman men who wrote the law, this principle seemed reasonable and natural. In many regards, it makes sense to me, too, perhaps because I have benefited from this norm throughout my life, including in my own pursuit of icebergs. In Greenland, I was able to go to the beach and collect ice like Kistaaraq Abelsen. I did not ask anyone, and no one stopped me. The iceberg was just there, and I was the first one to get to it. In Canada, too, I enjoyed the fruits of my server's labor after he hopped on his Jet Ski and scooped up a piece of ice floating in the waves off the coast of L'Anse aux Meadows.

In these instances, possession bestowed ownership. The growlers were ostensibly like the fox in *Pierson v. Post*: unowned and free for the taking until someone fully captured them. As every law student learns, however, this line-drawing is an arbitrary judgment. The judges could just as easily have decided that pursuit was enough to justify ownership, as is the case with a harpooned but not yet sub- dued whale. We do not have to treat icebergs like a member of the Canidae family. Nor do we have to rely on ancient Roman principles that can easily be distorted and abused. After all, when lines are drawn arbitrarily, some factors might be given more or less weight. Just think of the Europeans who inherited these laws and exploited the correlated idea of terra nullius, or "nobody's land," to justify the occupation of desired territories abroad, despite their knowledge that Indigenous peoples lived there.

Icebergs are a particularly tricky kind of *res*, or "thing." They are technically abiotic components of fragile ecosystems, but are more like living creatures than rocks since they are born and die. Icebergs are also itinerant. The frozen beasts roam the open ocean, moving in and out of waters claimed by neighboring countries. An iceberg calved from Sermeq Kujalleq in Ilulissat first swims through Disko Bay then travels north. After swirling around Baffin Bay for a year or more, floating in and out of Greenlandic and Canadian waters, it follows the Labrador Current along Iceberg Alley, past L'Anse aux Meadows, past St. John's, and into the unpredictable Grand Banks, just like the iceberg that sank the RMS *Titanic*. From there, the berg could continue its journey all the way to the Azores, or it might spontaneously disintegrate.

The Greenlandic schoolteacher Kistaaraq Abelsen, the waiter in L'Anse aux Meadows, the iceberg cowboy Ed Kean, Heiner Schwer from Berlin, and Abdulla Alshehhi could all theoretically capture the same iceberg. If icebergs are considered res nullius, the law of finders keepers governs the water to an extent. But is there a way to avoid having "losers" when we think about icebergs as property? Like earlier resource booms, the race for icebergs is largely unregulated. Few domestic laws address icebergs and no international legal instrument clarifies what speculators might do or must refrain from doing with the freshwater resource. In the past decade, only a handful of scholars have addressed the question of iceberg ownership. All identify the legal vacuum, but none make a substantive suggestion about how to fill it. Instead, some researchers have suggested how the United States might actually take advantage of the lack of clear laws. Nevertheless, most agree that icebergs are res nullius—that is, "nobody's thing." There have, however, been no attempts to define what kind of *res* icebergs are.

Although there is scant reference to icebergs in current laws, the icy masses exist in a complex web of international and domestic legal regulations thanks to their maritime setting and tendency to drift between borders. This tangle of laws is built on the United Nations Convention on the Law of the Sea (UNCLOS) and the set of treaties and agreements comprising the Antarctic Treaty System (ATS). Together, these international instruments establish the two overarching legal contexts in which iceberg harvesting may occur. UNCLOS defines the rights of nations to use the oceans and both living and nonliving resources in them. In addition to the European Union, 167 countries have joined UNCLOS. The United States remains a nonparty. However, because all nations are bound by customary international law, which closely mirrors UNCLOS, even non-signatories follow UNCLOS in practice. Moreover, the United States, Russia, Norway, Denmark, and Canada have committed to following the law of the sea in managing the Arctic. For much of the ocean, UNCLOS is thus the governing law. UNCLOS divides the ocean into maritime zones with distinct rights and regulations. Accordingly, the rules governing icebergs will change depending on where an iceberg is found.

The first zone is the territorial sea, which can extend up to twelve nautical miles from a coastal state's baseline, where the land meets the shore. Within this area, a state exercises near complete sovereignty and may set its own laws to regulate resources in the air, water, and seabed. It must allow innocent passage of foreign vessels through this water, but such vessels cannot take resources and still be considered innocent.

Beyond the territorial sea, coastal states may declare an exclusive economic zone (EEZ) that extends two hundred nautical miles from the baseline. EEZs comprise nearly one-third of the planet's oceans. In its EEZ, a coastal state has sovereign rights to exploit or conserve living and nonliving resources in the water column and underlying seabed, such as fish, oil, and natural gas. These rights, however, are accompanied by responsibilities. Coastal states must promote the optimum utilization of living resources with their EEZs. Theoretically, this means a country might have to allow another country to use its resources if it cannot meet this requirement. When it comes to nonliving resources like minerals and hydrocarbons in an EEZ, though, the state has no obligation to use or conserve them. This is why countries like Greenland, Canada, and Norway can presently make whatever rules they would like when it comes to icebergs in these waters.

◆

Iceberg harvesting in Canada is regulated by the provincial government of Newfoundland and Labrador, which includes icebergs in its Water Resources Act. Harvesters who plan to use icebergs for commercial purposes are required to obtain a license by paying a fee and supplying information to the government. The license limits how much water can be harvested and how it can be used and stipulates that it does not grant any right that might affect iceberg harvesting by residents of nearby communities for traditional personal domestic uses.

Licensees must follow several rules. They cannot use explosive material, and they cannot interfere with marine wildlife, fishing, or recreation activities. In fact, harvesting cannot be carried out "within visible distance from known locations frequented by tourists." The Canadian government also requires licensees to harvest only one iceberg at a time and to "identify or mark the iceberg" they intend to collect. So an iceberg in Newfoundland is not like a fox in New York—it does not need to be fully captured, just marked, to become property. Still, many of the details need to be worked out. For instance, Dr. Amir Ali Khan, manager of the Water Rights, Investigations, and Modelling Section of the Water Resources Management Division in the government of Newfoundland and Labrador, explained to me via email that the division has not enumerated what constitutes marking, as the number of users is currently low. "The condition has been incorporated to deal with a future situation where there could be more users with more potential for conflict," Dr. Khan elaborated. The provisional government is also prepared to limit the total number of licensees in case the number increases "to a point that there is the potential of conflict between the users or conflict with the tourism industry." But for now, there is no limit on the number of users who can harvest bergs. This includes both locals and outsiders. Foreign corporations could obtain a license, but bulk export would not be allowed. In addition to the commercial harvesters, Canadians also have a long-standing practice of collecting icebergs for personal use. This practice is unregulated and the icebergs are treated as res nullius; the first person to collect a bergy bit or growler can use it for themself.

Under the previous Greenlandic government, corporations could apply for similar licenses to harvest icebergs in the country's territorial waters. As Marlene Kongsted, head of the Ministry for Business and Trade, explained to me, the new government has suspended this strategy and is not accepting new requests to harvest water. Kongsted said that it may be an option in the future. As things stand, it is unclear if outsiders can harvest icebergs in Greenland for commercial purposes. Individuals in Greenland, though, can continue to harvest icebergs for their own use, just like Abelsen and her friends.

Norway has even fewer explicit laws on the books. When Jamal Qureshi had the idea to collect icebergs for Svalbarði, he simply contacted the environmental office of the Svalbard governor and shared his plan. The entrepreneur remembers the Norwegian civil servants' surprise, but they stated they had no objections to the plan and now Qureshi just has to keep the governor's office apprised of how many icebergs he is taking. Linda Karlstad filled in some details for me. While overlooking scenic Adventfjorden in a Scandinavian-chic atrium in Longyearbyen, the legal adviser with the Department for Environment Protection confirmed Qureshi's story and attempted to answer my questions about future concerns. Karlstad reported that there is no official threshold for what constitutes an appropriate number of icebergs that can be harvested. When I ask what would happen if several people wanted to collect icebergs, the trained jurist pulled at the sleeves of her red-and-white wool sweater. After some thought, she cited the Svalbard Environmental Protection Act from 2001, which is meant to preserve a virtually untouched environment in Svalbard. Practically, this means that iceberg harvesting cannot negatively impact the environment. The governor has not thought much about these issues, though, "because it has not been a problem." This is, of course, a common theme when it comes to iceberg harvesting. And, to be fair, the government in Svalbard probably has more pressing issue to resolve than speculate about future iceberg use. "Anyway," Karlstad told me, "I don't think it will happen that more people will want to harvest icebergs." For now, the handshake agreement between the entrepreneur and the government works well since Qureshi is personally invested in Svalbard and there is only one of him.

Even countries without formal iceberg-specific laws might control iceberg harvesting via analogous legal tools, particularly environmental laws. For instance, icebergs regularly float near Heard Island, an Australian External Territory that sits at fifty-three degrees south. The Australian government does not regulate water resources on the island, nor are there formal restrictions on taking water, but harvesting an iceberg in these territorial waters could implicate the Australian Environmental Protection and

Biodiversity Conservation Act. Because the island is a Common-
wealth Marine Reserve, extracting an iceberg from the area
could trigger a referral to the Department of Environment,
which would probably require an environmental impact assessment
and official government approval. So even if there are not official
iceberg laws on the books, certain domestic laws may offer some
guidance. The situation is similar under international law.

◆

Under UNCLOS, the high seas begin two hundred nautical miles
from the coastal baseline. All states, whether coastal or landlocked,
enjoy the freedom of the high seas. Despite popular perception, the
high seas are not completely lawless. UNCLOS requires states to
cooperate with each other in the management of living resources
in these global commons and encourages the use of regional fishery
management organizations. As a result, almost every nautical
mile of the high seas is governed by such organizations, like the
Indian Ocean Tuna Commission and the North Atlantic Salmon
Conservation Organization. Below the water column, beyond the
outer continental shelf of any coastal state, lies what is known as
"the Area," regulated by its own set of rules.

A separate set of treaties, known as the Antarctica Treaty
System (ATS), governs the globe south of the 60th parallel. To
guarantee peace while promoting scientific research, the ATS bans
military activity on the continent and suspends the determination
of the legal validity of territorial claims to Antarctica while prohibiting new claims from being made. Freezing the status quo has
proven an acceptable solution, in part, because there is no need to
define who owns what.

Because no signatory of the ATS can claim sovereignty over any
part of Antarctica, under UNCLOS the water meeting its coastline

is considered the high seas. Unlike other parts of the ocean, this high sea is governed by several agreements that provide policy mechanisms for environmental management, like the 1991 Protocol on Environmental Protection to the Antarctic Treaty (known as the Madrid Protocol). The Madrid Protocol obligates parties to plan and conduct activities in the Antarctic Treaty area, which includes the high seas south of sixty degrees, to "limit adverse impacts on the Antarctic environment." All mineral resource activities, unless for scientific research, are prohibited. More broadly, the Madrid Protocol requires parties to avoid activities that would result in "significant changes in the atmospheric, terrestrial (including aquatic), glacial or marine environments." At the 1989 Antarctica Treaty Consultative Meeting, iceberg harvesting was banned on the basis that it could have an adverse effect on the unique Antarctic environment.

For this reason, the thinkers behind POLEWATER, the UAE Iceberg Project, and the Southern Ice Project have decided to harvest icebergs above the 60th parallel. Although some glaciologists speculate that it will be more dangerous to harvest icebergs the longer they have been at sea, the law likely prevents iceberg harvesting closer to Antarctica. But, once icebergs cross that threshold, "they are free for the taking," Heiner Schwer informs me. That is, so long as the iceberg is not in the EEZ of a country with possessions in the Southern Ocean. Tiny, uninhabited Bouvet Island, for instance, belongs to Norway, which means that two hundred miles around it the seas are the exclusive economic zone of Norway. The same goes for the British Overseas Territory Gough Island and the dozens of other Antarctic islands.

In the rest of the world, the legal circumstances are more complicated. Consider the hypothetical iceberg that calved from Sermeq Kujalleq. An ordinary Greenlander or a tourist like me could probably capture it without any special permission. A foreign corporation, however, can no longer secure a license to harvest the iceberg in Greenlandic waters. That business might have to wait until the iceberg crosses midway through Baffin Bay, at which point it enters Canadian waters. Now Canada regulates who may take the iceberg. Depending on its path, the iceberg might float in and out of Canada's EEZ and the high seas, so maybe a harvesting license from the provincial government of Newfoundland and Labrador would be needed, or maybe anyone could take the iceberg. If the iceberg passes through the Grand Banks and into the broader Atlantic, then it would be fair game until floating into the waters around the Azores controlled by Portugal.

◆

Around the world, people treat icebergs like they are res nullius. Ed Kean thinks of growlers like deer—one just needs a license to hunt and then one can take whatever is subdued. Jamal Qureshi and the governor of Svalbard together decided that this resource could be owned by Qureshi if he captured it. And Kistaaraq Abelsen and her friends collect bergy bits that wash up on the beach for their own use. Icebergs, it seems, are free for the taking.

In a res nullius situation, as property becomes more valuable, individuals with access to the resource might act according to their own self-interest and not the common good. They might ignore the impacts of using the resource and disregard how other people might want to use it. Economists call this a common-pool resource problem, and they would consider icebergs rivalrous, nonexcludable goods. Everyone can access—or try to access—the resource and consume it, so icebergs are nonexcludable. And, because they perish after they are used, icebergs are considered rivalrous since they can only be consumed by a single user. In this regard, icebergs are like the commons—a shared resource in which everyone has an equal interest. When the commons are subject to a first come, first served regime, the so-called tragedy of the commons might result. Unco-ordinated, unchecked, and unregulated exploitation might cause the resource to be overused, the environment to be harmed in the process, and the benefits to be unequally distributed.

To be clear, Svalbarði, POLEWATER, and Iceberg Vodka, or any of the other extant or currently proposed iceberg ventures alone or even collectively, do not risk creating a tragedy of the commons. There are currently plenty of icebergs for people to use. The potential problems vis-à-vis the environment and the equitable distribution of this resource could result if any of these corporations grew large and greedy, or if a large number of harvesters flooded the ocean, or if a government decided to get in the game without regard for its neighbors' needs. What if iceberg harvesting becomes more affordable? Half a century ago, scientists laughed at Prince Mohamed Al-Faisal for suggesting that we might tow icebergs. Now it is possible. Maybe in another fifty years, harvesting icebergs will be the obvious way to collect freshwater.

The trouble with current plans to harvest icebergs is not the threat they individually pose to the environment, nor the possibility that they will control this resource at the exclusion of others. The problem is the principle they represent. Martin Riese, the water sommelier, disagrees with me, saying, "Some people might be saying that Jamal is a villain who is commoditizing a basic human right." He elaborates, "But they are just thinking about hydration. And if all they think of water is H$_2$O, they just don't know that something else is there." Dr. Olav Orheim, who has been involved in iceberg towing efforts for five decades, agrees that fairness "is not the main problem." He wants to first show that iceberg towing can be done, "so take icebergs where people can pay for them." From his perspective, the environmental harms are also protected by the matter of scale, and "even if it could come about, that is decades down the line." Most people I interview, in fact, dismiss the potential harms from iceberg harvesting, citing the small scope of an individual harvester in the wide ocean. Nevertheless, men like Qureshi and Schwer are normalizing assumptions and principles that may eventually prove harmful and difficult to reverse. They are paving the way for companies like Nestlé or Coca-Cola to stake a claim to this resource and pursue it with their deep coffers and profit-driven approach, potentially collecting the best bergs before those with humanitarian aims can reach them and doing so with little regard for the environment.

If icebergs are indeed res nullius, a number of unanswered questions remain, as the judges in *Post v. Pierson* made clear two centuries ago. How do icebergs come to be owned? Who should have the opportunity to own them? How many can one person, or corporation, or country own? Does it matter that icebergs swim through multiple territorial waters? In the case of icebergs, there is little guidance for how to treat these res nullius objects. The few legal scholars who have thought about icebergs as a fresh-water resource have declared a legal vacuum, meaning there are not preexisting rules about how to treat icebergs. But this does not mean we have to conceptualize icebergs in a vacuum. Several legal frameworks already exist that might apply, including treating icebergs as water, mineral resources, living creatures, found objects, or the common heritage of humankind. Admittedly, none makes perfect sense, but each offers a lesson about how we might think about icebergs.

◆

For obvious reasons, we might treat icebergs like water, a classic example of res nullius. But international water law is notoriously murky and mostly involves rules for navigation or energy production and not consumption of water itself. Plus, because icebergs move between territories, they most resemble international watercourses—like a glacier-fed tributary that moves through multiple countries—and the logics governing them, rather than laws concerning inland bodies of freshwater. The UN Watercourses Convention offers some guidance for how icebergs might be treated under international law if we were to consider them water. The convention, to which the United States is not a party, incorporates general principles from customary international law, including equitable and reasonable utilization, which requires us to consider the social and economic needs of the populations dependent on the watercourse, how their use might affect other states, and how the use might impact the environment, among other factors. The UN Watercourses Convention also directs states to consult each other and work cooperatively. States, in other words, are meant to allocate water fairly among each other and protect the ecosystems of international watercourses.

In the context of iceberg harvesting, this would seem to suggest the harvesting state would need to conduct an environmental impact report before the harvesting could occur and ensure that the resource is equitably used. Upstream users would have to heed the needs of downstream users. Greenland, for example, would have to mind Canadian uses. If the watercourse of an iceberg is conceptualized as ending in the high seas, then the needs of every country would need to be considered. Unfortunately, the convention does not provide much guidance on how to allocate water, though it states that "special regard" must be given to vital human needs. To date, the focus of this requirement has been on avoiding life-threatening dehydration. Under this reasoning, a country like South Africa might be entitled to icebergs over a country with sufficient water.

These ambiguities highlight one of the weaknesses of treating icebergs as international watercourses. It is uncertain which states might be considered co-riparian, and it is unclear which states should be cooperating, let alone how they should distribute the resource. Fortunately, there is no reason why we have to treat icebergs like other freshwater bodies of water. After all, clouds, like icebergs, are also water in a nonliquid state, and there is no legal expectation in international or domestic law that they be considered a freshwater resource.

◆

Thinking of icebergs as a mineral resource, like oil, might make better sense. The frozen blocks of ice are not replenishable and therefore do not require careful husbanding to ensure replenishment and environmental integrity. Each iceberg is a finite resource. And, technically, icebergs are minerals since they are naturally occurring solid inorganic substances. In this case, UNCLOS and the ATS offer pretty clear guidance.

If icebergs are treated like a nonliving resource, a state could exercise exclusive sovereignty over icebergs in its territorial waters and exclusive economic zone. Consequently, the state could harvest as many or as few as it liked. Setting aside technological limitations, a country like Greenland could also net every iceberg and sell them to wealthy corporations or countries, or, more realistically, invite them into Greenlandic waters to harvest the best bergs. In such a scenario, a wealthy state might end up with the lion's share of harvestable icebergs. Or icebergs might go unutilized.

Beyond the EEZ, UNCLOS prevents states from exploiting nonliving resources. As a result, icebergs in the high seas would be off-limits and left to melt. Treating icebergs like oil would therefore favor certain countries while disadvantaging others, and could lead to underexploitation in the high seas.

An iceberg floats in Disko Bay on the western coast of Greenland. *Courtesy of the author.*

A tabular iceberg towers above sea ice in McMurdo Sound in Antarctica. *Photograph by Peter Rejcek/ United States Antarctic Program.*

ABOVE: The International Ice Patrol was established in 1914 following the sinking of the RMS *Titanic*. It is funded by countries around the world and operated by the U.S. Coast Guard. *Emblem courtesy of the International Ice Patrol.* BELOW: Bernice Palmer snapped this photo of the iceberg that hit the RMS *Titanic* in the early hours of April 15, 1912 while safely aboard the RMS *Carpathia*. *Courtesy of the Division of Work and Industry, National Museum of American History, Smithsonian Institution.*

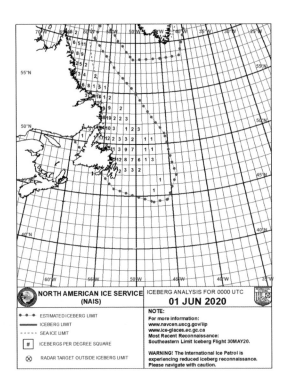

LEFT: An example of a Daily Iceberg Limit, the "warning product" the International Ice Patrol produces to keep the maritime community safe from icebergs. *Courtesy of the International Ice Patrol.* BELOW: A member of the International Ice Patrol monitors icebergs in the North Atlantic aboard a C-130 to create the Daily Iceberg Limit. *Courtesy of the U.S. Coast Guard. Courtesy of the Department of Defense.*

In June 1960, the International Ice Patrol conducted a number of tests to determine the best way to destroy icebergs. This 325-foot-long berg, floating in Newfoundland's Cape Bonavista Bay, withstood multiple blasts. *Courtesy of the Library of Congress.*

To speed the deterioration of "the mariner's ancient enemy," the International Ice Patrol also experimented with lamp black in 1960. Despite spreading 100 pounds of the carbon across an iceberg, little headway was made. *Courtesy of the Library of Congress.*

Members of the International Ice Patrol used a gasoline-powered ice drill to bore a hole to pocket a thermite charge atop an iceberg north of St. John's in 1960. *Courtesy of the Library of Congress.*

"Flame, steam, smoke and spiraling of ice bits erupt in a volcanic display of fireworks." So read the official description of the thermite tests. Despite the spectacle, only a small crater was left in the berg. *Courtesy of the Library of Congress.*

ABOVE AND BELOW: Icebergs float in Disko Bay on the western coast of Greenland. *Courtesy of the author.*

The Shapes of Icebergs

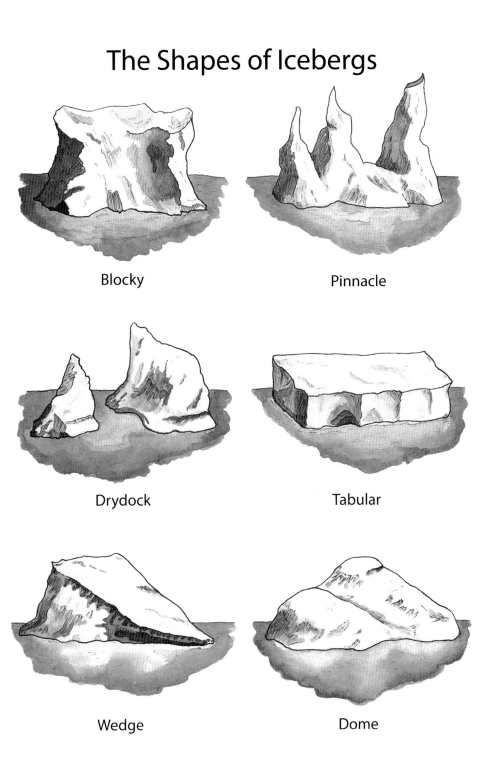

Blocky

Pinnacle

Drydock

Tabular

Wedge

Dome

Illustrations by Estelle Olivia.

THE ICE ISLANDS, seen the 9ᵗʰ of Janʸ 1773.

Captain Cook caused a sensation for harvesting icebergs in the Southern Ocean in the eighteenth century, depicted here by William Hodges. *The Ice Islands, Seen the 9th of Jan. 1773.* Engraving by B.T. Pouncy and W. Watts. Printed for W. Strahan and T. Cadell, 1777. *Courtesy of Beinecke Rare Book and Manuscript Library.*

Ed Kean harvests icebergs in Newfoundland, Canada. *Photograph by Veronique de Viguerie.*

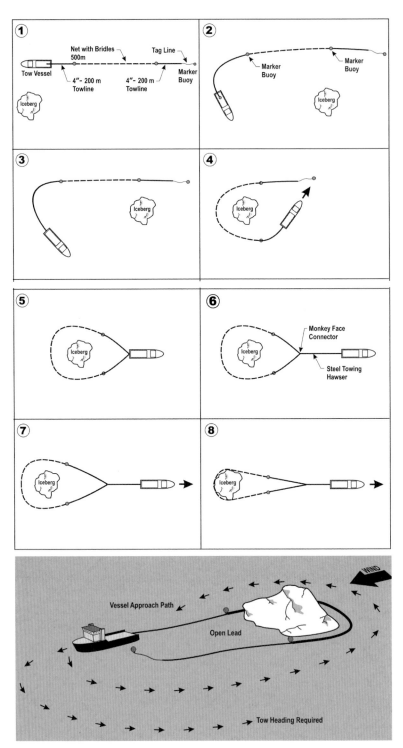

A step-by-step guide to towing icebergs produced by C-Core, a Newfoundland-based R&D company. *Courtesy of Freeman Ralph.*

BOTTOM: C-Core assists with an iceberg tow in Iceberg Alley. *Courtesy of Freeman Ralph.*

Frederic Edwin Church's monumental painting, *The Icebergs*, astounded viewers when it was debuted in 1861. *Oil on canvas, 64 ½ x 112 ½ inches. Courtesy of the Dallas Museum of Art, gift of Norma and Lamar Hunt.*

An iceberg stands in the Grand Palais for the Chanel Fall/Winter 2011 show during Paris Fashion Week. *Photograph by Michel Dufour.*

Harvested in the Arctic Ocean off the coast of the Norwegian island Svalbard, Svalbarði offers customers around the world the chance to taste pristine iceberg water. *Image courtesy of Jamal Qureshi.*

A tabular iceberg floats amid sea ice off the Larsen C ice shelf in Antarctica. *Photograph by Jeremy Harbeck/NASA.*

ABOVE: A rendering of POLEWATER's innovative suction buckets to keep icebergs steady while freshwater is harvested. *Courtesy of Heiner Schwer/POLEWATER.* BELOW: Towing an iceberg in the Southern Hemisphere will likely involve several steps. (1) Capture an iceberg north of 60° S. (2) Tow the berg into a helpful current and let nature take over. (3) Transfer the berg into a northward flowing current. (4) Anchor the iceberg in a suitable location to harvest the water. *Graphic by Estelle Olivia.*

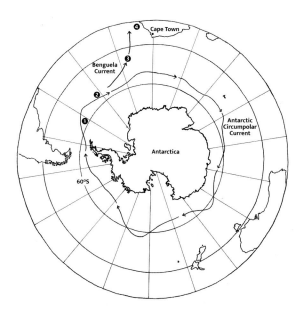

ABOVE: A mile-long iceberg calves in McMurdo Sound. *Photograph by Peter Rejcek/United States Antarctic Program.* BELOW: The largest recorded iceberg, Iceberg B-15, calved from Antarctica's Ross Ice Shelf in March 2000 and stretched more than 180 miles long. It broke into several pieces, the largest of which, B-15A, covered more than 2,400 square miles. Here, the northern edge of B-15A can be seen floating in the Ross Sea. *Photograph by Josh Landis/United States Antarctic Program.*

Despite their technical status, icebergs are not like other minerals. Icebergs move and do not wait around to be used. They will disappear into the ocean, becoming useless to the millions of people in desperate need of freshwater. Perhaps for this reason the Antarctic Treaty System has indicated that iceberg harvesting should not be considered a mineral activity, so there is a basis for eschewing handling them as such.

◆

Even though icebergs are nonliving resources, conceptualizing them as fish is logical since icebergs have a lifespan. Since there are so many icebergs in the ocean and we will have a never-ending supply in our lifetimes at least, icebergs might also be thought of as an infinite resource more like fish than oil. UNCLOS requires the harvest of non-straddling stocks in the high seas be for peaceful purposes, and be conducted in accordance with other international law rules, including environmental laws, and due regard to be given to the interests of other states. States must therefore cooperate with each other to conserve and manage these living resources by considering "relevant environmental and economic factors, including the special requirements of developing States."

Icebergs are most similar to anadromous fish, which are born inland and migrate to the sea. Like Pacific salmon, for example, icebergs might more specifically be considered straddling stocks, or fish that move between multiple territorial waters and/or the high seas.

The state where an anadromous stock originates has the primary interest in and responsibility for the resource under UNCLOS. As part of the requirement to conserve the stock, that state can set the optimum utilization, or a total allowable catch. In practice, this would allow the state to exclude other countries from using the resource. So UNCLOS also requires the state of origin to cooperate with other states that harvest the resource where the stock migrates through multiple EEZs. In the high seas, the state of origin must minimize economic harms and ensure that states that already participate in the harvest can continue to secure their "normal catch." If the state of origin sets an optimum utilization for harvesting icebergs, landlocked and geographically disadvantaged states in the same region or subregion have certain rights under UNCLOS, too. The terms of such states' participation must be established bilaterally through agreements and take into consideration, among others issues, the "nutritional needs" of the respective states. The coastal state must also "minimize detrimental effects" on harvesting communities and economic dislocation "in States whose nationals have habitually fished in the zone." In these cases, states must agree how to manage the resource to ensure optimum utilization. Most use regional fisheries management organizations, such as the North Atlantic Fisheries Organization or the Norwegian-Russian Fisheries Commission.

Under this model, a country like Greenland would have to consider the Canadians, and potentially others, who rely on icebergs that originate from Greenland. And countries that exploited Antarctic icebergs would have to protect the environment and work with each other to ensure everyone had access to the resource. But if the majority of icebergs never make it out of territorial waters in the Arctic, it might not be fair to treat icebergs like a straddling stock. Moreover, icebergs from different places behave differently. Icebergs that calve from Svalbard, for instance, might never make it to the exclusive economic zone of another country. This may result in the creation of different subspecies of icebergs, depending on their proclivities to travel along ocean currents. Apart from the administrative difficulties, this scheme would limit the states with an interest in icebergs to those states in the same region and those with an existing iceberg harvesting custom. Countries like the United Arab Emirates, South Africa, and the United States would likely not have any claim to the freshwater.

◆

The old maritime idea of flotsam could apply to icebergs, too. Historically, jetsam entailed sunken goods purposefully thrown overboard to save a ship. Flotsam, conversely, comprised goods that float on the seas that left a vessel through a natural action rather than by a deliberate act of the crew. It is considered abandoned when there are no assertions of ownership and it is "returned to the state of nature," as one US court pronounced. Flotsam is thus the equivalent to property like fish with no prior owner; it is res nullius. The reasoning behind the customary maritime law is to return lost property back into the stream of commerce to maximize its economic potential. Icebergs are naturally occurring derelicts that float on the seas; hence, if we think of a glacier or ice sheet as a ship, icebergs are a kind of flotsam.

Under UNCLOS, icebergs as such would be treated differently depending on where they are found. In a country's EEZ or territorial waters, flotsam is governed by the domestic laws of the state, regardless of where it came from or where it was going. In the high seas, UNCLOS would apply, which arguably requires flotsam to be preserved or disposed of for the benefit of humankind as a whole, with particular regard for the state or country of origin. Technically, the state of origin gets "preferential rights," though there is little consensus on the interpretation of this part of the convention. If icebergs are harvested in the high seas under this regime, they should be used for the benefit of everyone.

◆

Finally, we can learn lessons about how to treat icebergs by looking to resources designated the common heritage of humankind. Under UNCLOS, no state, no corporation, and no person can claim sovereignty over the resources in the Area. Instead, the rights in the resources are vested in humankind as a whole.

Although some countries urged commercial exploitation of the minerals lying deep under the ocean, others feared a first come, first served regime would launch a neocolonial scramble, favoring wealthy states with the technological capabilities to mine the resources. In response, the United Nations created the International Seabed Authority to administer and allocate rights to explore the deep seabed and exploit its resources, while protecting the marine environment, on behalf of humankind.

The business arm of the International Seabed Authority, known as the Enterprise, is meant to safeguard the interest of developing states while permitting economic development. To mine for polymetallic nodules beneath the deep ocean floor, for instance, nationals of a state party to UNCLOS must apply for an exploration license and then an exploitation license, which would grant the private company property in the minerals acquired. In its application, the private company requests a designated area to explore. It then must divide it into two parts of equal estimated commercial value—one is reserved for the Enterprise or developing states to operate with the assistance of the International Seabed Authority. The exploration licenses typically last fifteen years and cost around $500,000. If the contract is granted and commercial production is commenced, the contractor must pay an annual fixed fee of $1 million or pay a production charge, whichever is greater. For the first five years, the harvesting state need not make any payment; thereafter, it must contribute between 1 and 7 percent of annual production value. The Enterprise then equitably distributes the royalties, taking into account the interests and needs of developing states. The contractor must also devise practical training programs for the International Seabed Authority and developing states to engage in exploitation, and must share its technology to ensure that the Enterprise, and thus developing countries, can also engage in exploitation. The process of acquiring this property is far more involved than Post and Pierson's effort to claim a fox.

The treatment of the deep seabed as the common heritage of humankind has not been uncontroversial. The United States objected to the principle of equitably sharing the benefits derived from the use of deep seabed resources enshrined in UNCLOS. Many countries and corporations have also protested the requirement to share technology. Furthermore, there is little agreement about what constitutes equitable distribution of the funds raised from the licenses and the opportunities created. And the International Seabed Authority has been criticized as corrupt, untransparent, and favoring wealthy private sector actors over the well-being of the environment. Others criticize the designation of the deep seabed resources as "the common heritage of humankind," since the legal status has traditionally been used to prevent exploitation, as with the resources in Antarctica and the moon. For these reasons, the International Seabed Authority has not yet issued any exploitation licenses.

Icebergs are not technically in the Area, since they are on the surface of the ocean and not at or below the deep ocean seabed. And we do not want to freeze access to this freshwater resource, as has proven the case with the minerals regulated by the International Seabed Authority. But the principles undergirding the designation of the seabed as the common heritage of humanity, including environmental stewardship, the promotion of peace and solidarity, and encouraging distributive justice between nations, might be applied to a law governing icebergs.

◆

Just because icebergs are res nullius does not mean that harvesting icebergs must be unregulated. We ought to decide what kind of property icebergs are and what actions are required to transform icebergs from unowned wild beasts into private property. International law is replete with examples of different natural resources that are treated like different kinds of property with different rules governing their use and regulating who gets to benefit and how. There is no natural way to treat icebergs. Any rule, including having no rules at all, is an act of arbitrary line-drawing.

Some advocates of water justice might object to the characterization of icebergs as "nobody's thing." They contend that water belongs to all of us collectively. These activists believe it is wrong to commoditize water, to profit from its sale, and that governments should supply safe freshwater as a matter of human rights. The problem with Nick Sloane, Heiner Schwer, Abdulla Alshehhi, Ed Kean, and Jamal Qureshi from this perspective is that these men—perhaps there is something to be interrogated there—are operating on the principle that they can do what they would like with this resource without considering larger questions of need and fairness. They are normalizing the treatment of water as something that an individual can capture and sell, or choose to give away.

Yet, if this untapped resource is going to help solve the global freshwater shortage, we cannot say they belong to all of us. Simply arguing that water is part of a global commons does not tell us how water can be utilized. And, in the case of icebergs, if they are not harvested, they will disappear. It is not a resource we can wait around to use. We need to decide who gets to use icebergs—and how and how many—in a way that is just and equitable. It is technologically feasible to harvest and tow icebergs great distances. It might be possible to raise money to fund the effort. But that does not ensure that the people who need freshwater will get what is melted from the polar giants. We need a legal system in place that ensures that finders can be keepers, but that losers are not weepers.

If icebergs are to be used equitably across the globe, in light of Indigenous water justice and environmental best practices, it is essential to define the legal status of icebergs before custom does so for us. In international law, the customary behavior of states is elevated to the status of legal obligation if there is sufficient state practice and *opinio juris*. The first element considers the duration, consistency, repetition, and generality of the behavior—the material fact. Historically, more politically powerful states have played a larger role in the establishment of customs. For instance, the former Soviet Union and the United States shaped much of space law. Custom is not legally binding, however, until there is wide acceptance, or *opinio juris*, the second element. This is the "belief that [a] practice is rendered obligatory by the existence of a rule of law requiring it."

Subsequently developed laws may also take custom and traditional practices into consideration, like UNCLOS, which aimed to codify existing state practice and customary international law. If private corporations start taking icebergs wherever and whenever they like, this could be the norm in the future, no matter the goals of someone like Nick Sloane, the needs of downstream iceberg cowboys, or local interests in Greenland. If a handful of powerful countries begin harvesting icebergs, they could dictate the law of icebergs in the future. And legal scholars have shown that customary international laws tend to be economically inefficient, do not respect Indigenous rights, and benefit wealthy states. Because water is essential to human life and icebergs have the potential to do so much good, a clear idea should be formed before custom can shape international law. Even if it is a long shot that the International Court of Justice would recognize traditional iceberg law, and even if a country like the United States would ignore the practice of other states, the stakes are too high to risk the consequences of doing nothing.

Conclusion
Making the Dream
a Reality

Icebergs could save our planet. More than ten thousand icebergs flood the Southern Ocean every year and up to forty thousand calve into the Arctic. They contain enough freshwater to save millions of people for years to come. Yet, throughout the world, people struggle to find safe freshwater. According to the World Health Organization, two billion people use a drinking water source contaminated with feces, and every two minutes a child under five dies from an illness linked to poor water and sanitation. More than a quarter of the world's population live in water-stressed countries. And our struggles are only going to grow worse. As climate change exacerbates existing access to clean freshwater, the demand for water will increase 55 percent by 2050 as the global population surges. Fatefully, our warming planet has concurrently created a solution in the form of tidy parcels of perfectly pure frozen freshwater floating in the oceans. We just need to collect them. Otherwise, if we let these icebergs roam the ocean frontier and die their natural deaths, the precious resource they contain will mingle with salty seawater, rendering them useless for the purposes of alleviating the mounting freshwater crisis. Rescuing humanity, however, will require more than just the herculean feat of harvesting the ice.

According to an old cowboy adage apocryphally attributed to Mark Twain, "Whiskey is for drinking, water is for fighting," meaning that valuable resources, when scarce and ownership is disputed, can easily lead to violence. Scholars of water history—there is even a *Water Conflict Chronology*—have documented hundreds of violent water-related fights around the world, from Bangladesh to Israel to East Timor to Serbia to Peru to the United States. As they analyze the situation, opposing views about the control and distribution of a shared resource like water easily lead to conflict. Dr. Ismail Serageldin, former chairperson of the World Commission on Water and vice president of the World Bank, has predicted that "the wars of the 21st century will be fought over water." The way out, he declared, is to change our approach to managing the precious resource.

Only a handful of people are paying attention to icebergs now, but as the race for icebergs heats up, financial desires, cultural practices, humanitarian needs, and environmental concerns may come into direct conflict. There is no guarantee that those who want to use icebergs to solve the water crisis will succeed. Although a war over icebergs is unlikely, the res nullius object could result in a resource race without rules in place. An unchecked contest over icebergs could lead to the water they contain ending up in the hands of the wealthy elite and devastating environmental degradation.

Property rights can help avoid fights over common-pool resources. The rights do not necessarily have to be formalized in law. The Nobel Prize–winning economist Elinor Ostrom showed that local communities can manage valuable resources themselves without turning to the state for administration and without creating private property schemes. She even uses several

examples involving water to make her point, including the irrigation mechanisms between Valencia and Alicante in Spain and
the *zanjera* irrigation community in the Philippines, to show that
public planning is not always necessary. Rather, bottom-up rules
can emerge and still ensure the peaceful, economically efficient,
and sustainable management of resources.

To be successful, these local forms of self-government must
have clear rules that are shared by the community. Another
essential element, according to Ostrom, is the establishment
of democratic decision-making that involves all users of the
resource. Consequently, the communities that succeed in self-
managing common-pool resources are generally small. In the
case of icebergs, the community that is interested in the resource
is the entire planet. And some parties, like Greenlanders and
Newfoundlanders, may feel like they have a greater interest in
and claim to the resource. Other people, such as South Africans,
may feel like they have a special interest because they need freshwater more than Greenlanders, who have an abundance of water.
Or maybe Emiratis feel entitled to the resource because they can
pay the most for it.

The world is too big and our interests are too diverse to
expect such bottom-up rules to emerge to govern icebergs. Jamal
Qureshi's goals are too different from Heiner Schwer's ambitions
and Abdulla Alshehhi's plans, let alone those of a conglomerate
motivated to tap into this resource solely for profit. Plus, even if
informal norms could regulate iceberg harvesting, I am not sure
we would want them to, since the first actors to capture icebergs
in international waters will likely prefer a finders keepers, losers
weepers regime. Instead, we need top-down laws that can achieve
the balancing act of rewarding entrepreneurs for assuming risk,
safeguarding Indigenous rights, and supplying water for those in
need, all while protecting the environment.

◆

Ensuring icebergs are used in a way that most benefits the planet without destroying it requires a global solution in the form of international laws. Of course, this proclamation assumes there is agreement about what is considered "beneficial" for people, and that is rarely the case. If you do not care if the neediest people have access to icebergs, then maybe you prefer the legal status quo. If your primary objective is to support Arctic Indigenous peoples, you may want a law that permits Greenland to exercise exclusive sovereignty over this resource. If you believe risk-taking entrepreneurs should be rewarded for recognizing an overlooked resource, then perhaps you desire a law that keeps icebergs res nullius in the high seas.

From my perspective, the challenge is to develop a regulatory framework that helps humanitarians like Heiner Schwer without disenfranchising other stakeholders like Kistaaraq Abelsen and Ed Kean, and without giving too much power to corporations focused exclusively on profits. I do not have the answer for what, exactly, the law should be, but there are many principles already undergirding rules governing shared resources in the ocean that can be productively applied to icebergs.

Since there is little point using icebergs to save the planet if we destroy Earth in the process, environmental regulations must be part of any regulatory scheme we create to govern iceberg harvesting. As a matter of scale, towing a few icebergs itself will not greatly harm the polar environments from which they are taken. However, we should monitor how many icebergs we remove to ensure we do not unleash unintended consequences, like destroying carbon sinks that are created by phytoplankton that are nourished by icebergs.

It seems like we can worry less about local ecosystems in South Africa, the United Arab Emirates, and wherever else icebergs might turn up. Although some damage will be inevitable, it may be minimal and worth it for the sake of supplying a life necessity. On this basis, many countries permit offshore wind farms that disrupt the seabed in order to supply electricity to people onshore. If anchoring an iceberg causes similar damage, it might be easy to justify. Ideally, an environmental impact study would identify all of the risks specific to a targeted ecosystem, but the variables involved in iceberg towing may be too great to get a clear picture before we actually accomplish the feat.

Fortunately, ecosystems have the ability to adjust. Dr. Waleed Abdalati reminded me that "nature is good at handling slow perturbations." So if we are going to introduce an iceberg to Cape Town, we should do it slowly to let nature adjust. This would also give us the time to monitor the effects, the NASA scientist reasons. Dr. Claire L. Parkinson agrees with Dr. Abdalati about taking environmental change slowly. She says that "our success will be linked to how gently we proceed into the future." Icebergs may be foreign invaders in their new ecosystems, but the natural world can be good at welcoming strangers.

On principle, Dr. Abdalati is not opposed to all geoengineering schemes. "I'm not as troubled by the thought of towing icebergs to places in need of freshwater," he says, "as I am by the idea of pouring aerosols into the atmosphere to cool the earth," referring to the popular idea to combat climate change. As he reasons, towing icebergs is not adding anything new to the environment; it is just rearranging it. "And there is the possibility of something really good coming out of it."

Anyway, it is a bit spurious to reject iceberg towing as more or less natural than what is happening now. Without human involvement, icebergs would not travel all the way to Cape Town or Fujairah. But without humans, we would not have this many icebergs, either, since humans are responsible for the climate change that is precipitating calving events in the Arctic and the

Antarctic. Dr. Abdalati reminds me that humans are part of the ecosystem, too. Addressing our needs could be considered part of a natural process.

Environmental philosophers might argue that this is an overly anthropocentric view of the world. Species besides humans rely on freshwater, and nature itself might even have rights to be free from human interventions. Some scholars advocate for human extinction as the best means of saving the planet and the rich variety of life it contains. If this is one's position, then we should probably not rearrange the globe to use icebergs for the sake of humankind. If one thinks that we should try to save people in need of freshwater, even if it comes at the cost of other living and nonliving parts of our world, then iceberg towing may be a good solution, so long as we proceed gently, which means we need rules.

◆

Since water is always situated and places us in relationship with each other, the laws we craft to govern icebergs should consider who currently relies on icebergs—whether for backyard barbecues in Ilulissat, sightseeing in Newfoundland, or commerce in Longyearbyen—and recognize that icebergs are part of a global, interconnected ocean, to which any person, from any corner of the world, has a claim. The law should also acknowledge the debt we owe inhabitants of the Northern Hemisphere, particularly Indigenous peoples, for safeguarding this valuable resource.

Hypothetically, Greenland could sell all of the best icebergs calving off the island to a rich company or country. In such a scenario, people in places like Cape Town would likely not benefit from icebergs and residents of Fujairah could enjoy an emerald oasis. To prevent such a near-monopoly on icebergs north of the equator, we can compel states of iceberg origins to consider the needs and uses of downstream users. By applying the requirements applied to anadromous fish under UNCLOS, a new international law of icebergs could ensure the resource is distributed to those who have traditionally used it. Greenland and Canada could create the equivalent of a regional fishery management organization, like the Norwegian-Russian Fisheries Commission. Yet, I am hesitant to curb hard-won Greenlandic sovereignty, especially because Greenlanders have helped keep this resource clean and pure. We could adopt Abelsen's idea to apply to icebergs the regulations governing the sale of sealskin products. The European Union only permits trade in sealskins from hunts conducted by Inuit and other Indigenous communities, which ensures the practice is ethical and sustainable while supporting Indigenous peoples. Perhaps we should require that trade in icebergs in either hemisphere only be conducted if Indigenous partners are involved in the business. This would balance Indigenous water justice with the practices of others who rely on icebergs and still create room for entrepreneurs to pursue profits. It would also be a modest step toward compensating Indigenous peoples around the world for the colonial losses they have endured when it comes to water.

To ensure icebergs actually get used, we cannot remove financial incentives for those who are willing to assume the risk of trying to tow these giants. This means we should treat icebergs like flotsam and not minerals in the high seas. Any corporation or country attempting to harvest the treasure in the high seas should feel free to capture the res nullius object, subject to UNCLOS obligations not to damage the environment and to pursue the activity for peaceful purposes. Harvesters should also be required to secure insurance, like the folks at the Southern Ice Project, who have indicated that their operation would be insured by Lloyd's of London in case an iceberg breaks apart and endangers other ships or vulnerable parts of the polar or local ecosystem.

However, we need to make certain that the neediest people benefit from icebergs. The team at POLEWATER is committed to giving away iceberg water for free, and for a price, Nick Sloane is willing to drag a block to a country desperate for water. There is no guarantee that another corporation would do that. We may further aspire to avoid a situation where a private corporation has the power to decide who should get emergency freshwater, as much as I admire Schwer and Schwarzer. While there is no agreement on what constitutes equity in international law, I believe we can devise a workable system as we go. This will require principles like those undergirding the notion of the common heritage of humankind under UNCLOS, which seeks to ensure that all peoples can utilize the resources of the ocean floor and not simply those with the capital and technological capacity. In a new law of icebergs, we can require the sharing of economic benefits and knowledge gained through iceberg harvesting similar to the Enterprise in the Deep Seabed Authority.

The best way to regulate icebergs while fulfilling these values is to establish an intergovernmental body responsible for authorizing and controlling iceberg harvesting. Ideally this would be done through a UN convention like the International Seabed Authority—after all, the United Nations already has over thirty organizations working on water-related issues. Experts from around the world could craft the specific rules and regulations governing icebergs without leaving private corporations—no matter how good their intentions—to determine the law of icebergs for the future. And UNCLOS has already created a framework that applies throughout the world's oceans onto which icebergs can be mapped as a new resource.

An International Iceberg Authority is not such a farfetched idea. In the *Histories*, Herodotus shows us how having a common enemy, the Persians, helped unite the Greek city-states. Hostile icebergs already brought the world together once after the RMS *Titanic* sank. Today the International Ice Patrol is actively expanding the scope of its international collaboration. In addition to working with nations adjacent to the North Atlantic, the International Ice Patrol is sharing information and data with Australia and Argentina to develop better iceberg models for the Southern Hemisphere and to manage the global threat posed by icebergs. If we continue to think about icebergs as a foe that requires coordination to best combat, we may find intrinsic motivation to collaborate between nations and diverse interest groups.

An International Iceberg Authority would bring together scientists, private sector leaders, humanitarians, and policymakers at the United Nations. We need to mobilize to take action sooner than later, before men like Sloane, Alshehhi, and Schwer take to the seas. And we need these big thinkers to commit to working with the international community and not go at this project alone.

◆

Even if we develop a legal system delineating property rights in icebergs, it will not save the planet if people do not think about icebergs as a solution to the water crisis. In his 1798 ballad *The Rime of the Ancient Mariner*, Samuel Taylor Coleridge penned the famous line "water, water every where / nor any drop to drink." The English poet was likely inspired by James Cook's second voyage, during which the captain and Georg Forster, driven by their desperation for freshwater, turned to icebergs. It has taken centuries for people outside the Arctic to learn Cook's lesson, and still today many people do not realize that icebergs contain valuable freshwater. The ocean, from their perspective, is full of unpotable water.

Once more people recognize that the Arctic and the Antarctic are chock full of freshwater, we need to inspire them to invest in towing and melting icebergs. This task, as Sloane, Alshehhi, and

Schwer know, is difficult. One strategy, effectively deployed by Qureshi, is to market icebergs as precious luxuries. Arcticism can sell icebergs at high prices. While this may help defray harvesting costs and turn a profit, it will not save the planet. To do that, we need to shift the way that icebergs are perceived.

A good first step is communicating the ubiquity of icebergs. Coleridge's poem also includes an accurate description of crackling and growling emerald green icebergs: "The ice was here, the ice was there / the ice was all around." But that stanza has not entered the common vernacular. Few people outside of the Arctic get to see icebergs, which makes the sparkling gems feel rare and exotic. This is why tourists flock to Newfoundland and Labrador for a chance to see the stunning sculptures glide through the ocean and why people in West Virginia thought iceberg water seemed inherently fancy and extravagant.

To convince governments and investors that icebergs are worth pursuing, we need to talk about icebergs as an untapped resource and not as mystical objects. We need to think about icebergs less as sparkling jewels earmarked for elite consumption and more as an accidentally persevered reservoir to which everyone is entitled, perhaps especially the poorest among us. The success of Qureshi's business model highlights the conceptual ground we need to cover to make this shift. But such a change is possible—after all, it was once widely believed that drinking icebergs was dangerous. We can change the way we think.

Icebergs themselves may help. The blank white surface is a projection site for our beliefs and fantasies, both sacred and profane. Instead of being a depressing sign of global warming and a signal of future disaster, what if icebergs were recast as a symbol for our dream of international cooperation and humanitarianism? A reminder that we are all on this planet together and share a need for its resources. Can we see the sheen of icebergs as a glimmer of hope?

Paradoxically, harvesting icebergs will help shift our perspective. The first time an iceberg turns up in Fujairah or Cape Town, it will appear as a marvel. As more icebergs arrive and the water they release is given away, the ice will seem less exotic and less elite. It will be understood as the saving grace our planet has frozen for us and preserved from our own pollution. But only if icebergs are actually utilized to help those in need. Unfortunately, there is nothing in international or domestic law that currently requires icebergs be used in this fashion. Only our cultural and political beliefs can spur this action.

◆

Icebergs can be collected and melted to provide lifesaving water to people around the planet if the right coalition of visionaries, scientists, engineers, lawyers, and diplomats unites to ensure the neediest and not just wealthy consumers benefit. A new international agreement must spell out who gets to use icebergs while reserving rights for landlocked nations and Indigenous peoples. Much like how the United Nations Convention on the Law of the Sea created the International Seabed Authority to govern mining activities beneath the high seas, the international community will need to develop a centralized organization to control iceberg utilization, force technology sharing, and delineate what constitutes equitable distribution. Pooling our knowledge will further result in a clearer picture of the environmental costs and enable us to make a better-informed decision about how harvesting this freshwater will influence climate change and vice versa.

As more and more icebergs are launched into the oceans as the planet warms, now is the time to come together. We should unify our objectives before private interests exploit the legal vacuum and before they establish a customary practice that could eventually inform any international law created to govern the resource. Otherwise it might be too late to ensure that the resource is equitably used and the hunt for icebergs does not spiral out of control.

Even if tens of thousands of icebergs are calved each year, not all of them will be harvestable. Some will be better than others. Without regulations in place, the best icebergs—the biggest ones with the right shape and most life left, in easy to access locations and well positioned to tow—will likely be harvested by companies with the resources to go after them first and fastest, and not necessarily by humanitarians or countries in need of water. And even if there were hypothetically enough good icebergs to go around, from an environmental perspective we would not want them all harvested. The consequences could be dire. We need a system to figure out who gets to use icebergs and how.

If we do not work in coordination, another future may unfold. The Cold Rush may become a free-for-all if more and more people start trying to tow icebergs. Unregulated resource booms often result in interpersonal and interstate conflict. Private corporations and wealthy states profit while those in acute need often come out on the bottom. In almost every case where there are no governing rules, Mother Earth pays the biggest price of all. My conversations with Kistaaraq Abelsen, Commander Marcus Hirschberg, and Dr. Ellen Stone Mosley-Thompson make me believe we can avoid this fate. And the international coalition built by Nick Sloane is just one indication that people from around the world are willing to work together to realize the dream of harvesting icebergs.

Epilogue

I am sitting two hundred meters above the Persian Gulf in an opulent restaurant cantilevered from the iconic Burj Al Arab Hotel in Dubai. After years of visiting the Arctic and traveling to cold places, Jordan planned a warm-weather vacation. I personally prefer the soggy bogs of L'Anse aux Meadows and the sparsity of Longyearbyen to the pristine sands and metropolitanism of Dubai, but I am not one to complain, especially if it means I can enjoy high tea in the $1 billion sail-shaped landmark.

Below me, a series of artificial islands stretches into the sea to form a sprawling palm tree so large it can be seen from space. When the archipelago was first announced, many doubted the feasibility of completing such a large-scale project. Since, it has been hailed as a miracle of planning, engineering, and investment. The Palm Jumeirah, which purportedly cost $12 billion to construct, now houses luxury resorts and multimillion-dollar villas spread across its fronds. Behind me, the Burj Khalifa, the world's tallest building, towers over an indoor ski resort and a thirty-acre fountain that can spray 22,000 gallons of water in the air in an instant.

Since there is no iceberg water on the menu—yet—I sip champagne and imagine how Emiratis might react to a berg floating off the coast of such an extravagant skyline. Though the one-hundred-million-ton icebergs Alshehhi hopes to drag from Antarctica to Fujairah, just a ninety-minute drive away, weigh more than the Burj Khalifa and are taller than the Burj Al Arab, I am not so sure locals would be as astonished as the simulated crowd the promotional video depicts.

Dubai is a monument to what humans can achieve and the control we are able to exert over the natural world. It is also a testament to the devastating human and environmental costs of these achievements if we do not carefully think about our values as we strive for greatness. The rapid growth of Dubai and the lavish lifestyles it supports requires large quantities of freshwater, which is sparse in the desert nation. The United Arab Emirates consequently relies on desalination plants, which produce tons of carbon dioxide and leave the sea saltier than ever. The Gulf's salinity level has now risen to a level that endangers local marine life and the nation has one of the largest carbon footprints per capita in the world. On land, the city has been built on the backs of men and women working in conditions that have routinely been condemned by Human Rights Watch and other international organizations. Many have had their passports confiscated, are paid a pittance, and toil in deadly conditions, including the 2,500 laborers who protested their plight as they erected the Burj Khalifa. The glittering metropolis is consequently disparaged by some as no more than a glamorous mirage masking misery.

Just because humans can do great things does not necessarily mean that we should. Mind-bending feats like fabricating a ski slope in the Arabian Desert, materializing islands in the Persian Gulf, and building a tower so tall it takes 110,000 tons of concrete and 55,000 tons of steel rebar have consequences. These consequences can be easy to overlook, especially for those of us lucky enough to comfortably sit in luxurious settings at floor-to-ceiling windows that make us feel like we have a clear-eyed view of the scene.

For that reason, it is important to think critically about what we do with icebergs now, before we are enchanted by their charms and stunned by our own ingenuity. Harvesting icebergs could be the next great human triumph. If we can change the way we think about the sparkling ice as a resource for the world, create a legal framework, and invest in towing bergs for people in need and not just wealthy consumers, we can ensure the frozen freshwater is equitably shared and exploited in a safe manner. But if we take the next steps without clear rules, or if greedy actors pursue this resource without concern for the environment or human suffering, towing icebergs might harm the planet rather than help it.

Dante's *Divine Comedy* begins with the protagonist lost in a dark wood, assailed by beasts representing self-indulgence, maliciousness, and violence. As Dante sinks lower and lower, Virgil shows up to guide the Italian poet through hell so he might learn to recognize sin. At the conclusion of the *Inferno*, Virgil must say goodbye, and Dante continues his journey. Eventually, Beatrice leads Dante through heaven, where he achieves a new understanding of the divine. Perhaps regrettably, I am no Beatrice. I cannot precisely reveal how icebergs can save our planet. But I hope I have helped illuminate the pitfalls and helped set us on a path to achieve the ultimate goal.

Pursuing icebergs is especially attractive since other solutions to the growing global freshwater shortage have repeatedly proven ineffective, costly, and bad for the environment. Removing salt and other minerals from undrinkable seawater has been touted as an answer to our woes for decades. Around the world, ten trillion gallons of drinking water each year come from desalination plants, and the industry is continuing to grow. Existing infrastructure, however, relies on fossil fuels, threatens marine life by sucking water into intake pipes, pollutes the ocean with too much salt, and is prohibitively expensive for many of the places that most need freshwater. Reclaiming wastewater to supplement potable water supplies is another option. Technologies like reverse osmosis, ultraviolet disinfection, and activated carbon filtration can remove pathogens from sewage and purify the water. The capital of Namibia, Windhoek, has relied on such recycled water for more than fifty years. Water experts from Singapore, Australia, and the United States have studied the system there and brought back the idea to use at home. Wide-scale implementation, however, has been limited by regulatory standards, difficulties in identifying certain contaminants, monitoring costs, and psychological blocks that prevent consumers from enthusiastically embracing the final product. More recently, investors have poured millions into hyrdo-panels that use a fan to pull air through hygroscopic material that traps water. Such modular panels have been used in Puerto Rico to provide water after a hurricane and in an orphanage in Lebanon, as well as in American cities like Phoenix, Arizona. Critics worry the panels cannot be scaled to provide enough water at cost-effective rates, since each panel meets just over 10 percent of one person's daily water needs. Of course, the possibility of towing icebergs does not mean we should not also continue to improve these freshwater sourcing alternatives. Correspondingly, the existence of desalination and reclamation plants does not negate the need to seriously consider iceberg harvesting.

To tow icebergs, we will have to tap into everything they represent. The frozen crystals have been seen as divine gifts and exotic emissaries from the past and faraway places. They are deadly and best avoided, but lifesaving for thirsty captains brave enough to approach them. Icebergs are like migrating beasts but are actually abiotic components of complex ecosystems. They are remarkably ubiquitous and free for the taking. Paradoxically, icebergs are all things at once, which is why the twinkling forms have long inspired flights of fancy and why iceberg towing has been considered a joke. The forms are too full with cultural meaning to easily make sense of in our minds. Trying to manipulate icebergs, conceptually or physically, seems a hubristic endeavor.

This density of meaning can play to our advantage, however, by spurring us into action. Icebergs are undeniably beautiful, but not such rare and special gems that they should be earmarked for the elite alone. They are fortuitous gifts of perfectly packaged pure freshwater for the whole planet, preserved in the Northern Hemisphere thanks to Indigenous peoples and in the Southern Hemisphere thanks to international cooperation and a commitment to protect the environment. We must remember that icebergs are dangerous, but not use this memory to consider icebergs unharvestable. Rather, we should see icebergs as a reminder of the lethal consequences of uncoordinated human activity. Fortunately, icebergs have already brought the world together. These ice mountains are enmeshed in our art, literature, myths, fashion, politics, commerce, laws, and ecosystems. They are part of our history and part of our future. We ought to embrace the visions icebergs inspire and let ourselves dream that we can tow them around the world rather than worry about being duped by a chimera.

Nearly fifty years ago, Prince Mohamed Al-Faisal dared to share his dream in Ames, Iowa. At the time, his proclamation that we could tow an iceberg to the Middle East in three to five years was premature. It might not be now. Brilliant scientists, engineers, and entrepreneurs have continued to work on the ambitious project. It is time for us to catch up and formally develop the legal and philosophical principles we want to govern this valuable resource.

With the exception of those of you interested enough to go on this journey with me, few people have thought of icebergs as a freshwater source. This creates a powerful opportunity. In his *History of the Peloponnesian War,* Thucydides already recognized that "[m]ost people . . . will not take trouble in finding out the truth, but are more inclined to accept the first story they hear." We need to believe iceberg towing is more than a joke, otherwise it will be difficult to raise the funds to accomplish the feat. Or we might not be prepared when it does happen. If we want to see icebergs in Fujairah and Cape Town, we should decide now. We need to tell a story about icebergs saving the planet.

The future of icebergs also depends on the regulatory framework we will collectively create, the environmental policies we enact, and the ability of icebergs to bring us together rather than pull us apart. Iceberg harvesting will create exciting opportunities in the years to come. The ice could transform the desert around Fujairah into a lush oasis and might stun visitors to Cape Town. Other dreamers have already designed modular offshore farms built around icebergs to allow plants, fed by a stream of freshwater melting from the ice, to grow on the ocean. Because we cannot know what the future has in store, we must prepare for it as best we can. Climate change will continue to create terrible difficulties and worsen inequalities on Earth. The ice trapped in glaciers and ice caps will continue to melt at unprecedented rates. If we invest in icebergs, change how we see them, share them, and protect the environment in the process, then maybe we have a solution to the freshwater crisis exacerbated by the global warming we have caused. Icebergs may just be a silver lining.

Acknowledgments

This book would not have been possible without the many people who took the time to generously share their perspectives, experiences, and knowledge with me. In particular, I would like to thank Kistaaraq Abelsen, Waleed Abdalati, Doug Benn, Alan Condron, Anna Crawford, Tim Cronin, Mike Hicks, Marcus Hirschberg, Linda Karlstad, Ed Kean, Amir Ali Khan, Marlene Kongsted, David Meyers, John Mortensen, Ellen Mosley-Thompson, Claire L. Parkinson, Louise O'Riordan, Olav Orheim, Santiago de la Peña, Jamal Qureshi, Freeman Ralph, Martin Riese, Jill Klein Rone, Timm Schwarzer, Heiner Schwer, Nick Sloane, Anja Sørenson, and Arthur von Wiesenberger.

I am eternally grateful to my agents Justin Broukaert and Todd Shuster at Aevitas Creative Management for helping me shape this project and for their wise counsel through every stage of the process. Likewise, I am indebted to the entire team at Pegasus Books and especially Jessica Case for her encouragement, understanding, and guidance as we navigated a global pandemic and the birth of my children during the completion of this manuscript. Estelle Olivia created exquisite graphics, Maria Fernandez produced beautiful design, and Jessica LeTourneur

Bax provided impeccable copyediting. Any mistakes that remain are entirely of my own creation.

For providing valuable feedback, tips, inspiration, and logistical support, I thank Maria Ackrén, Athena Angelos, Kris Barrett, Anna Berman, Mara Brown, Katra Byram, Peter Brooks, Gary Hayward, Mohamed Helal, Sabine Höhler, Robert Holub, Irina Ikonsky, Anne Karasinski, Lauren Kostinas, Katie Kotol, May Mergenthaler, Natascha Miller, Andrew Norman, Kay Peterson, Paul Reitter, Andrew Robinson, Larry Rosen, Silvan Salvisberg, Marc Spindelman, Robin Stephenson, Tanvi Solanki, Ebbe Volquardsen, and Andrea Westermann. I have further benefitted from my colleagues at the Ohio State University who offered intelligent criticism, and I was vitally helped by the staff at the Byrd Polar and Climate Research Center and the Ohio State University Libraries. I owe an especial thanks to my parents, Lee Stephenson Birkhold and Richard Birkhold, and my brothers, Alexander Birkhold and Adam Birkhold, for counting among my best readers and supporters. My brilliant husband, Jordan Elkind, has been my fellow adventurer, editor, and champion. He makes everything better. Finally, I thank Bjorn Privin and Ursula Ojaniemi for motivating me to remain optimistic about the future and what we might achieve.

SOURCES

LEGAL SOURCES

Act of June 15, 2001, No. 29, Relating to the Protection of the
Environment in Svalbard.

An Act Respecting the Control and Management of Water Resource
in the Province, SNL2002, Chapter W-4.01.

The Antarctic Treaty, opened for signature December 1, 1959, 402
UNTS 71, entered into force June 23, 1961.

Declaration on the Rights of Indigenous Peoples. G.A. Resolution
61/295, UN Doc A/RES/61/295, September 13, 2007.

Environment Protection and Biodiversity Conservation Act 1999.
No. 91, 1999.

Guidelines for Firearms and Protection and Scaring Devices against
Polar Bears. Adopted and entered into force July 20, 2021,
pursuant to Act No. 7 of April 20, 2018.

The Ilulissat Declaration. May 28, 2008.

Indigenous Peoples Kyoto Water Declaration. March 2003.

Lov om Svalbard. Passed July 17, 1925, entered into force August 14,
1925.

North Sea Continental Shelf (Ger. v. Den.; Ger. v. Neth.), Judgment,
I.C.J. Reports, 3, February 20, 1969.

Pierson v. Post, 3 Cai. R. 175 (1805).

Protocol on Environmental Protection to the Antarctic Treaty.
Opened for signature October 4, 1991, 2941 UNTS 3, entered
into force January 14, 1998.

Recommendation XV-21 (ATCM XV–Paris, 1989).

R.M.S. Titanic, Inc. v. Wrecked & Abandoned Vessel, 435 F.3d 521 (2006).

United Nations Convention on the Law of the Non-Navigational Uses of International Watercourses. G.A. Res. 51/229, at 1, UN Doc. A/RES/51/206, May 21, 1997, entered into force August 17, 2014.

United Nations Convention on the Law of the Sea, opened for signature December 10, 1982, 1833 UNTS 396, entered into force November 16, 1994.

WEBSITES

https://www.arcticbiodiversity.is
https://www.deutscher-meerespreis.de/2015_en.html
https://finewaters.com
http://icebergs.world
https://www.martin-riese.com
https://www.polewater.com
https://svalbardi.com
https://www.worldwater.org/water-conflict/
https://www.youtube.com/watch?v=CdC3HPNjc94

BOOKS, ARTICLES, AND REPORTS

2030 Water Resources Group Annual Report 2021, 10-11, https://2030wrg.org/wp-content/uploads/2021/12/WRG -Annual-Report_2021_RV_Sprds_2_10_22.pdf

Åhrén, Mattias. *Indigenous Peoples' Status in the International Legal System*. Oxford: Oxford University Press, 2016.

Alighieri, Dante. *The Divine Comedy of Dante Alighieri*, Inferno. Translated by Allen Mandelbaum. New York: Bantam Books, 2004.

Allen, Paddy, and Jenny Ridley. "The 'Cold Rush': Industrialisation in the Arctic." *Guardian*, July 5, 2011.

Argus. April 7, 1855.

Armstrong, James C., and Nigel A. Worden. "The Slaves, 1652– 1834." In *The Shaping of South African Society, 1652–1840*, edited by Richard Elphick and Hermann Giliomee, 109–83. Middletown, CT: Wesleyan University Press, 1979.

Baldwin, Derek. "Trial Run for UAE Iceberg Project in 2019." *Gulf News*, July 7, 2018. https://gulfnews.com/uae/environment/trial-run-for-uae-iceberg-project-in-2019-1.2244996.

Bamberger, Werner. "Icebergs Defy New Coast Guard Assaults." *New York Times*, July 17, 1960.

———. "International Ice Patrol Ends; Coast Guard Spend Busy Season; Attempts to Destroy Bergs by Bombing Not Successful—Data Gathered to Improve Detection by Radar." *New York Times*, July 26, 1959.

Barlow, Maude. *Blue Covenant: The Global Water Crisis and the Coming Battle for the Right to Water.* New York: The New Press, 2007.

Barry, Roger G., and Eileen A. Hall-McKim. *Polar Environments and Global Change* Cambridge: Cambridge University Press, 2018.

Bartholomaus, T. C., C. F. Larsen, S. O'Neel, and M. E. West. "Calving Seismicity from Iceberg–Sea Surface Interactions." *Journal of Geophysical Research* (December 22, 2012). doi:10.1029/2012JF002513.

Becker, Elizabeth. *Overbooked: The Exploding Business of Travel and Tourism.* New York: Simon & Schuster, 2013.

Benner, Susanne, Gregor Lax, Paul J. Crutzen, Ulrich Pöschl, Jos Lelieveld, and Hans Günter Brauch, eds., *Paul J. Crutzen and the Anthropocene: A New Epoch in Earth's History.* Cham: Springer, 2021.

Biggs, Grant R. *Icebergs: Their Science and Links to Global Change.* Cambridge: Cambridge University Press, 2015.

Boelens, Rutgerd, Jeroen Vos, and Tom Perrault. "Introduction: The Multiple Challenges and Layers of Water Justice Struggles." In *Water Justice*, edited by Rutgerd Boelens, Jeroen Vos, and Tom Perrault, 1–32. Cambridge: Cambridge University Press, 2018.

Boretti, Alberto, and Lorenzo Rosa. "Reassessing the Projections of the World Water Development Report." *npj Clean Water*, 2 no. 15 (2019). https://www.nature.com/articles/s41545-019-0039-9.

Borrelli-Persson, Laird. "The Big Chill: Revisiting Chanel's Fall 2010 Iceberg Show." *Vogue*, December 27, 2018. https://www.vogue.com/article/karl-lagerfeld-imported-an-iceberg-from-sweden-for-his-fall-2010-chanel-show.

Boston Globe. September 3, 1914.

Bowler, Peter J. "Geographical Distribution in the *Origin of Species.*" In *The Cambridge Companion to the "Origin of Species,"* edited by Michael Ruse and Robert J. Richards, 153–72. Cambridge: Cambridge University Press, 2009.

"The Brains behind the Costa Concordia Salvage." *Local*, September 17, 2013. https://www.thelocal.it/20130917/the-mastermind -behind-the-concordia-salvage-operation.

Breum, Martin. *Cold Rush: The Astonishing True Story of the New Quest for the Polar North.* Montreal: McGill-Queen's University Press, 2018.

———. "Her er den egentlige forskel på dansk og grønlandsk syn på fremtiden." *Altinget*, January 9, 2019. https://www.altinget.dk /arktis/artikel/martin-breum-her-er-den-egentlige-forskel-paa -dansk-og-groenlandsk-syn-paa-fremtiden.

Broekhoff, Derik, Michael Gillenwater, Tani Colbert-Sangree, and Patrick Cage. "Securing Climate Benefit: A Guide to Using Carbon Offsets." Stockholm Environment Institute & Greenhouse Gas Management Institute, 2019. www .offsetguide.org.

Brown, Mark. "How to Tow a Building-Sized Iceberg." *Wired*, August 10, 2011, https://www.wired.com/2011/08/ iceberg-towing-drinking-water.

Butler, Daniel Allen. *The Other Side of the Night: The* Carpathia, *the* Californian, *and the Night the* Titanic *Was Lost.* Philadelphia: Casemate, 2009.

Campbell, Gordon. *Norse America: The Story of a Founding Myth.* Oxford: Oxford University Press, 2021.

Caravella, Kristi Denise, and Jocilyn Danise Martinez. "Water Quality and Quantity: Globalization." In *Managing Water Resources and Hydrological Systems*, edited by Brian D. Fath and Sven Erok Jorgensen, 265–76. Boca Raton, FL: CRC Press, 2021.

Carr, Gerald L. *Frederic Edwin Church: The Icebergs.* Dallas: Dallas Museum of Fine Arts, 1980.

Carroll, Siobhan. *An Empire of Air and Water: Uncolonizable Space in the British Imagination, 1750–1850.* Philadelphia: University of Pennsylvania Press, 2015.

Carter, Andrew, Teal R. Riley, Claus-Dieter Hillenbrand, and Martin Rittner. "Widespread Antarctic Glaciation during the Late Eocene." *Earth and Planetary Science Letters* 458 (2017): 49–57.

Cecco, Leyland. "Spirited Away: Canadian Thieves Steal More Than $9,000 in Iceberg Water." *Guardian*, February 14, 2019. https://www.theguardian.com/world/2019/feb/14/canada -iceberg-water-theft-vodka.

Clark, Christopher. *The Sleepwalkers: How Europe Went to War in 1914.* New York: HarperCollins, 2013.

Coleridge, Samuel Taylor. *The Major Works*. Edited by H. J. Jackson.
    Oxford: Oxford University Press, 1985.
Cook, James. *The Journals of Captain Cook*. Edited by Philip Edwards.
    London: Penguin, 2003.
"Costa Concordia Capsizing Costs over $2 Billion for Owners."
    Reuters, July 2, 2014. https://www.reuters.com/article/italy
    -concordia-costs/costa-concordia-capsizing-costs-over-2-billion
    -for-owners-idUSL6N0PH0EO20140706.
Crowley, Roger. *Conquerors: How Portugal Forged the First Global
    Empire*. New York: Random House, 2015.
Darwin, Charles. *The Origin of Species*. Oxford: Oxford University
    Press, 1998.
Dawoud, Mohamed A., and Mohamed M. Al Mulla. "Environmental
    Impacts of Seawater Desalination: Arabian Gul Case Study."
    International Journal of Environment and Sustainability 1, no. 3
    (2012): 22–37.
Dig Deep and US Water Alliance. *Closing the Water Access Gap in the
    United States: A National Action Plan*, 2019, 12, https://www
    .digdeep.org/close-the-water-gap.
The Digest of Justinian. Vol. 4. Translated and edited by Alan Watson.
    Philadelphia: University of Pennsylvania Press, 1985.
Dodds, Klaus. *Ice: Nature and Culture*. London: Reaktion Books, 2018.
Douglas, Elizabeth. "Towns Sell Their Public Water Systems—and
    Come to Regret It." *Washington Post*, July 8, 2017. https://www
    .washingtonpost.com/national/health-science/towns-sell-their
    -public-water-systems--and-come-to-regret-it/2017/07/07/
    6ec5b8d6-4bc6-11e7-bc1b-fddbd8359dee_story.html.
Doyle, Alister. "Too Much Salt: Water Desalination Plants Harm
    Environment: U.N." Reuters, January 14, 2019. https://www
    .reuters.com/article/us-environment-brine/too-much-salt-water
    -desalination-plants-harm-environment-u-n-idUSKCN1P81PX.
Durfee, Mary, and Rachael Lorna Johnstone. *Arctic Governance in a
    Changing World*. Lanham, MD: Rowman & Littlefield, 2019.
Ellickson, Robert C. *Order without Law: How Neighbors Settle Disputes*.
    Cambridge, MA: Harvard University Press, 1991.
"Entombed in an Iceberg for Centuries." *Cincinnati Enquirer*,
    March 26, 1922.
European Environment Agency. *European Maritime Transport
    Environmental Report*. Luxembourg: Publication Office of
    the European Union, 2021. https://www.eea.europa.eu/
    publications/maritime-transport.

Fantini, Emanuele. "An Introduction to the Human Right to Water: Law, Politics, and Beyond." *WIREs Water* 7, no. 2 (2020). https://wires.onlinelibrary.wiley.com/doi/10.1002/wat2.1405.

Fassnacht, Martin Philipp, Nikolaus Kluge, and Henning Mohr. "Pricing Luxury Brands: Specificities, Conceptualization and Performance Impact." *Marketing: ZFP—Journal of Research and Management* 35, no. 2 (2013): 104–17.

Fishman, Charles. "Message in a Bottle." Fast Company, July 1, 2007. https://www.fastcompany.com/59971/message-bottle.

Flora, Janne. *Wandering Spirits: Loneliness and Longing in Greenland.* Chicago: University of Chicago Press, 2019.

Forrest, Craig. *International Law and the Protection of Cultural Heritage.* Abingdon, UK: Routledge, 2010.

Forster, Georg. *A Voyage Round the World.* Vol. 1. Edited by Nicholas Thomas and Oliver Berghof. Honolulu: University of Hawaiʻi Press, 2000.

Foster, John Wilson. Titanic: *Culture and Calamity.* Vancouver: Belcouver Press, 2016.

"A Genius in Bedford." *Scientific American*, August 22, 1863.

Gentleman, Amelia. "Coca-Cola Urged to Close an Indian Plant to Save Water." *New York Times*, January 16, 2008. https://www.nytimes.com/2008/01/16/business/16coke.html.

Geon, Bryan S. "A Right to Ice? The Application of International and National Water Laws to the Acquisition of Iceberg Rights." *Michigan Journal of International Law* 19 (1997): 277–99.

Gertner, Jon. *The Ice at the End of the World: An Epic Journey into Greenland's Buried Past and Our Perilous Future.* New York: Random House, 2020.

Glick, David M., Jillian L. Goldfarb, Wendy Heiger-Bernays, and Douglas L. Kriner. "Public Knowledge, Contaminant Concerns, and Support for Recycled Water in the United States." *Resources, Conservation and Recycling* 150 (2019). https://www.sciencedirect.com/science/article/pii/S0921344919303143.

Gomez, Jim. "China Coast Guard Uses Water Cannon against Philippine Boats." Associated Press, November 18, 2021. https://apnews.com/article/china-united-states-philippines-manila-south-china-sea-9fe8af0a7ae2386e058e99a7c5c33243.

Gornitz, Vivien. *Vanishing Ice: Glaciers, Ice Sheets, and Rising Seas.* New York: Columbia University Press, 2019.

Gosnell, Mariana. *Ice: The Nature, the History, and the Uses of an Astonishing Substance.* New York: Knopf, 2011.

Goss, Michael. *Lost at Sea: Ghost Ships and Other Mysteries.* New York: Prometheus Books, 1994.

Gothe-Snape, Jackson, and Emma Machan. "Why a Middle Eastern Business Thirsty for Water Can't Just Tow an Iceberg from Antarctica." *ABC News,* August 14, 2019, https://www.abc.net.au/news/2019-08-14/why-a-middle-eastern-business-cant-just-tow-antarctica-iceberg/11318638.

Gould, Tom. "Selling Water at $150/m3 to the World's Poorest People—with Billionaire Backing." *Atmospheric Water Generation* 21, no. 5 (May 21, 2020). https://www.global waterintel.com/global-water-intelligence-magazine/21/5/general/selling-water-at-150-m3-to-the-world-s-poorest-people-with-billionaire-backing.

Graugaard, Naja Dyrendom. "'Without Seals, There Are No Greenlanders,' Colonial and Postcolonial Narratives of Sustainability and Inuit Seal Hunting." In *The Politics of Sustainability in the Arctic: Reconfiguring Identity, Space, and Time,* edited by Jeppe Stransbjerg and Ulrik Pram Gad, 74–93. New York: Routledge, 2019.

Greenberg, James. "Greenland's 'Nuummioq' a Stunning Triumph." Reuters, January 25, 2010. https://www.reuters.com/article/us-film-nuummioq/greenlands-nuummioq-a-stunning-triumph-idUSTRE60P0EA20100126.

Gronholt-Pedersen, Jacob. "Left-Wing Warty Wins Greenland Election, Opposes Big Mining Project." Reuters, April 7, 2021. https://www.reuters.com/article/us-greenland-election/left-wing-party-wins-greenland-election-opposes-big-mining-project-idUSKBN2BU0V1.

Gross, Daniel A. "Recycling Sewage into Drinking Water Is No Big Deal. They've Been Doing It in Namibia for 50 Years." *The World,* December 15, 2016. https://theworld.org/stories/2016-12-15/recycling-sewage-drinking-water-no-big-deal-theyve-been-doing-it-namibia-50-years.

"The Guiding Star." *Sydney Morning Herald,* September 10, 1855.

Hardin, Garret. "The Tragedy of the Commons." *Science* 162, no. 3859 (1968): 1243–48.

Harkin, Michael E., and David Rich Lewis. "Introduction." In *Native Americans and the Environment: Perspectives on the Ecological Indian,* edited by Michael E. Harkin and David Rich Lewis, xix–xxxiv. Lincoln: University of Nebraska Press, 2007.

Harris, Michael. *Lament for an Ocean: The Collapse of the Atlantic Cod Fishery.* Toronto: McCelland & Stewart, 2013.

Hartwig, Lana D., Francis Markham, and Sue Jackson. "Benchmarking Indigenous Water Holdings in the Murray-Darling Basin: A Crucial Step towards Developing Water Rights Targets for Australia." *Australasian Journal of Water Resources* 25, no. 2 (2021): 98–110.

Hauptmann, Gerhart. *Atlantis.* Translated by Adele and Thomas Seltzer. New York: B. W. Huebsch, 1912.

"Hauptmann's Prize." *New York Times,* November 16, 1912.

Heese, Hans Friedrich. "Cape of Good Hope? Meeting Place of Unwilling Migrants from Africa, Asia, and Indigenous People." *Insights of Anthropology* 4, no. 1 (2020): 268–79.

Hemming, Henry. *The Ingenious Mr. Pyke: Inventor, Fugitive, Spy.* New York: PublicAffairs, 2015.

Herodotus. *The Histories,* translated by Robin Waterfield. Oxford: Oxford University Press, 2008.

"High Seas: Little *Titanic.*" *Time,* February 9, 1959.

Hoff, Mary. "Envision 2050: The Future of Oceans," Ensia, May 28, 2015. https://ensia.com/features/envision-2050-the-future-of-oceans.

Human Rights Watch. "United Arab Emirates, Events of 2021." https://www.hrw.org/world-report/2022/country-chapters/united-arab-emirates.

Hummels, David. "Transportation Costs and International Trade in Second Era of Globalization." *Journal of Economic Perspectives* 21, no. 3 (2007): 131–54.

Hundley, Norris. *Water and the West: The Colorado River Compact and the Politics of Water in the American West.* Berkeley: University of California Press, 2009.

Hünefeldt, Thomas, and Annika Schlitte. "Introduction: Situatedness and Place." In *Situatedness and Place: Multidisciplinary Perspective on the Spacio-Temporal Contingency of Human Life,* edited by Thomas Hünefeldt and Annika Schlitte, 1–18. Cham: Springer, 2018.

Husseiny, A. A., ed. *Iceberg Utilization: Proceedings of the First International Conference and Workshop on Iceberg Utilization for Fresh Water Production, Weather Modification and Other Applications held at Iowa State University, Ames, Iowa, USA, October 2–6, 1977.* New York: Pergamon Press, 1978.

Ibrahim, Hamed D., Pengfei Xue, and Elfatih A. B. Eltahir, "Multiple Salinity Equilibria and Resilience of Persian/Arabian Gulf

Basin Salinity to Brine Discharge." Frontiers in Marine Science, July 10, 2020, https://doi.org/10.3389/fmars.2020.00573.

"Iceberg in Near-Miss with Floating Oil Platform off Newfoundland." *Hamilton Spectator*, March 30, 2017. https://www.thespec.com/news/canada/2017/03/30/iceberg-in-near-miss-with-floating-oil-platform-off-newfoundland.html.

Intagliata, Christopher. "Greenland Is Melting Faster Than Any Time in Past 12,000 Years." Scientific American, October 3, 2020. https://www.scientificamerican.com/podcast/episode/greenland-is-melting-faster-than-any-time-in-past-12-000-years.

International Convention of the Safety of Life at Sea. *International Conference on Safety of Life at Sea: Messages from the President.* Washington, DC: Government Printing Office, 1914.

Iowa State University Yearbook. *The Bomb.* Ames: Bomb Publication Board, 1977.

Jack, Malcom. *To the Fairest Cape: European Encounters in the Cape of Good Hope* Lewisburg, PA: Bucknell University Press, 2019.

Jaeckel, Aline L. *The International Seabed Authority and the Precautionary Principle: Balancing Deep Seabed Mineral Mining and Marine Environmental Protection.* Leiden: Brill Nijhoff, 2017.

Jamail, Dahr. *The End of Ice: Bearing Witness and Finding Meaning in the Path of Climate Disruption.* New York: The New Press, 2019.

Jensen, Ole G. *The Culture of Greenland in Glimpses.* Nuuk: Milik Publishing, 2007.

Johnson, Robert Irwin. *Guardians of the Sea: History of the United States Coast Guard, 1915 to the Present.* Annapolis, MD: Naval Institute Press, 1987.

Junger, Sebastian. *The Perfect Storm: A True Story of Men against the Sea.* New York: W. W. Norton, 1997.

"Just the Tip of the Iceberg . . ." *Sydney Morning Herald*, March 29, 2003. https://www.smh.com.au/national/just-tip-of-the-iceberg-20030329-gdgiee.html.

Kadıoğlu, Uluç. "Taken Hostage in the UAE." *Harvard International Review*, July 29, 2022. https://hir.harvard.edu/taken-hostage-in-the-uae.

Kaiser, Gisela. *Parched: The Cape Town Drought Story.* Cham: Springer International Publishing, 2021.

Kakande, Yasin. *Slave States: The Practice of Kafala in the Gulf Arab Region.* Winchester, UK: Zero Books, 2015.

Kaplan, Sarah. "Snow Is Beautiful. Please Don't Eat It." *Washington Post*, January 22, 2016. https://www.washingtonpost.com/news

/morning-mix/wp/2016/01/22/snow-is-beautiful-please-dont
-eat-it.

Kavenna, Joanna. *The Ice Museum: In Search of the Lost Land of Thule.*
New York: Penguin, 2007.

Kibel, Paul Stanton. "Damage to Fisheries by Dams: The Interplay
between International Water Law and International Fisheries
Law." *UCLA Journal of International Law & Foreign Affairs* 121
(2017): 121–50.

Kneeland, Douglas E. "An Alaskan Iceberg Upstages a Saudi Prince at
Conference in Iowa." *New York Times*, October 7, 1977.

Kuhn, Eric, and John Fleck. *Science Be Damned: How Ignoring
Inconvenient Science Drained the Colorado River.* Tucson:
University of Arizona Press, 2019.

Leahy, Stephen, and Katherine Purvis. "Peak Salt: Is the Desalination
Dream Over for the Gulf States?" *Guardian*, September 29,
2016.

Lewis, Cory. "Iceberg Harvesting: Suggesting a Federal Regulatory
Regime for a New Freshwater Source." *Boston College
Environmental Affairs Law Review* 42 (2015): 439–71.

Lewis, Simon L., and Mark A. Maslin. "Defining the Anthropocene."
Nature 519 (2015): 171–80.

Low, Michael Christopher. "Desert Dreams of Drinking the
Sea, Consumed by the Cold War: Transnational Flows of
Desalination and Energy from the Pacific to the Persian Gulf."
Environment and History 7, no. 2 (2020): 145–74.

Lyell, Charles. *Principles of Geology.* Vol. 1. London: John Murray,
1830.

Mackley, Jude S. *The Legend of St. Brendan: A Comparative Study of the
Latin and Anglo-Norman Versions.* Leiden: Brill, 2008.

Madrigal, Alexis C. "The Many Failures and Few Successes of
Zany Iceberg Towing Schemes," *Atlantic*, August 10, 2011,
https://www.theatlantic.com/technology/archive/2011/08/the
-many-failures-and-few-successes-of-zany-iceberg-towing
-schemes/243364.

Mambra, Shamseer. "5 Biggest Oil Platforms in the World." Marine
Insight, March 9, 2022. https://www.marineinsight.com/
offshore/5-biggest-oil-platforms-in-the-world.

Mascha, Michael. *Fine Waters: A Connoisseur's Guide to the World's Most
Distinctive Bottled Waters.* Philadelphia: Quirk Books, 2006.

McCarthy, Joe, and Erica Sánchez. "This Clean Water Solution
Actually Produces Trillions of Gallons of Toxic Sludge

Worldwide." Global Citizen, January 15, 2019. https://www
.globalcitizen.org/en/content/desalination-water-toxic-sludge.

McLynn, Frank. *Captain Cook: Master of the Seas*. New Haven, CT:
Yale University Press, 2011.

McQuigge, Michelle. "Author's Fictional Titan Hit Iceberg Years
before the *Titanic* Sank." *Canadian Press*, April 12, 2012.

McVeigh, Karen. "Seabed Regulator Accused of Deciding Deep
Sea's Future 'Behind Closed Doors.'" *Guardian*, April 1, 2022.
https://www.theguardian.com/environment/2022/apr/01/
worlds-seabed-regulator-accused-of-reckless-failings-over
-deep-sea-mining.

Melville, Herman. *Moby-Dick; Or, The Whale*. New York: Penguin, 2003.

———. *Redburn*. New York: Random House, 2002.

Mikkelsen, Naja, and Torsten Ingerslev, eds. *Nomination of the Ilulissat
Icefjord for Inclusion in the World Heritage List*. Copenhagen:
Geological Survey of Denmark and Greenland, 2002. https://
whc.unesco.org/uploads/nominations/1149.pdf.

Miles, Vincent. *The Lost Hero of Cape Cod: Captain Asa Eldridge and
the Maritime Trade That Shaped America*. Yarmouth Port, MA:
Historical Society of Old Yarmouth, 2015.

Mival, Kenneth N. "Re-Assessing the Disappearance of the Clipper
'Guiding Star.'" *Great Circle* 37, no. 2 (2015): 16–39.

Mogul, Fred. "Offshore Wind May Help the Planet—but Will It Hurt
Whales." NPR, December 5, 2019. https://www.npr.org/2019/
12/05/782694371/offshore-wind-may-help-the-planet-but-will
-it-hurt-whales.

Molony, Senan. Titanic: *Why She Collided, Why She Sank, Why She
Should Never Have Sailed*. Cork: Mercier Press, 2019.

Morales, Laura. "Many Native Americans Can't Get Clean Water,
Report Finds." NPR, November 18, 2019.

Morrow, David. *Values in Climate Policy*. London: Rowman &
Littlefield, 2020.

Morton, Oliver. *The Planet Remade: How Geoengineering Could Change
the World*. Princeton, NJ: Princeton University Press, 2017.

Mull, Andrea. "How Fancy Water Bottles Became a 21st-Century Status
Symbol." *Atlantic*, February 12, 2019. https://www.theatlantic
.com/health/archive/2019/02/luxury-water-bottles/582595.

Müller, Christian, et al. "Singing Icebergs." *Science* 310, no. 5727
(2005): 1299–1303.

Navigatio Sancti Brendani Abbatis. Translated by John O'Meara
and Jonathan Wooding. In *The Voyage of Saint Brendan:*

Representative Versions of the Legend in English Translation,
edited by W. R. J. Barron and Glyn S. Burgess, 13–64. Exeter:
University of Exeter Press, 2002.

The New Oxford Annotated Bible with Apocrypha. Oxford: Oxford
University Press, 2018.

Notz, Dirk, and Julienne Stroeve. "Observed Arctic Sea-Ice Loss
Directly Follows Anthropogenic CO_2 Emission." *Science* 354,
no. 6313 (2016): 747–50.

O'Connor, Joe. "Chasing Cold Cash: How Icebergs Became the Field
of Dreams for Believers and Schemers." *Financial Post*, March 21,
2019. https://business.financialpost.com/entrepreneur/chasing
-cold-cash-how-icebergs-became-the-field-of-dreams-for
-believers-and-schemers.

O'Connor, Ralph, and Anne O'Connor. *Icelandic Histories and
Romances.* 2nd ed. Stroud: Tempus, 2006).

O'Donnell, Santiago. "The Iceberg Cometh: Drought: Towing
a Chuck of Glacier into the Harbor Is the 5th of 8 Water
Shortage Solutions the Council Plans to Look at in a $175,000
Study." *Los Angeles Times*, June 6, 1990.

Oldach, Eliza, Helen Killeen, Priya Shukla, Ellie Brauer, Nicholas
Carter, Jennifer Fields, Alexandra Thomsen, Cassidy Cooper,
Leah Mellinger, Kaiwen Wang, Carl Hendrickson, Anna
Neumann, Pernille Sporon Bøving, and Nann Fangue.
"Managed and Unmanaged Whale Mortality in the California
Current Ecosystem." *Marine Policy* 140 (June 2022). https://
www.sciencedirect.com/science/article/pii/S0308597X
22000860#bbib4.

Olivera, Oscar. *Cochabamba! Water War in Bolivia.* Cambridge, MA:
South End Press, 2004.

O'Neill-Yates, Chris. "Investigation Shows Husky's Decision Not
to Move Out of Iceberg's Path in Close Call 'Economically
Driven.'" CBC News, February 26, 2018. https://www.cbc
.ca/news/canada/newfoundland-labrador/husky-iceberg
-suspension-searose-fpso-suspension-cnlopb-offshore
-1.4551362.

Onishi, Norimitsu, and Somini Sengupta. "In South Africa, Facing
'Day Zero' with No Water." *New York Times*, January 30, 2018.

Organisation for Economic Co-operation and Development. *OECD
Environmental Outlook to 2050: The Consequences of Inaction*,
2012. https://read.oecd-ilibrary.org/environment/oecd
-environmental-outlook-to-2050_9789264122246-en.

Ostrom, Elinor. *Governing the Commons: The Evolution of Institutions for Collective Action*. Cambridge: Cambridge University Press, 2015.

Parisa, Ariya A., et al. "Role of Snow and Cold Environment in the Fate and Effects of Nanoparticles and Select Organic Pollutants from Gasoline Engine Exhaust." *Environmental Sciences: Processes and Impacts* 18, no. 2 (2016): 153–296.

Parker, Laura. "Microplastics Have Moved into Virtually Every Crevice on Earth." *National Geographic*, August 7, 2020. https://www.nationalgeographic.com/science/article/microplastics-in-virtually-every-crevice-on-earth.

Payoyo, Peter Bautista. *Cries of the Sea: World Inequality, Sustainable Development and the Common Heritage of Humanity*. The Hague: Kluwer Law International, 1997.

Pearse, Guy. *The Greenwash Effect: Corporate Deception, Celebrity Environmentalists, and What Big Business Isn't Telling You about Their Green Products and Brands*. New York: Skyhorse Publishing, 2014.

Plumwood, Val. *Environmental Culture: The Ecological Crisis of Reason*. New York: Routledge, 2002.

Potgieter, Thean. "South Africa: The Blue Economy Experience." In *The Blue Economy in Sub-Saharan Africa: Working for a Sustainable Future*, edited by Donald L. Sparks, 115–40. New York: Routledge, 2021.

Pringle, John. "A Discourse upon Some Late Improvements of the Means for Preserving the Health of Mariners." Delivered at the Anniversary Meeting of the Royal Society, November 30, 1776. In *A Voyage towards the South Pole and Round the World*, vol. 2, by James Cook. London: Strahan and Cadell, 1777.

Qadir, Manzoor, and Vladimir Smakhtin. "Where the Water Is." *Project Syndicate*, May 17, 2018. https://www.project-syndicate.org/commentary/tapping-unconventional-freshwater-sources-by-manzoor-qadir-and-vladimir-smakhtin-2018-05.

Raab, Jennifer. "'Precisely These Objects': Frederic Church and the Culture of Detail." *Art Bulletin* 95, no. 4 (2013): 587–96.

Raina, Aditi, Yogendra Gurung, and Bhim Suwal. "Equity Impacts of Informal Private Water Markets: Case of Kathmandu Valley." *Water Policy* 22 (2020): 189–204.

Rhind, William. *Elements of Geology and Physical Geography*. Edinburgh: Fraser & Co. J. Anderson, 1837.

Rink, Hinrich. *Tales and Traditions of the Eskimo with a Sketch of their Habits, Religion, Language and Other Peculiarities*. Translated by Robert Brown. Cambridge: Cambridge University Press, 2014.

Roberts, Andrew. *Churchill: Walking with Destiny*. New York: Penguin, 2018.

Robertson, Morgan. *Futility*. New York: M. F. Mansfield, 1898.

Robison, Jason, Barbara A. Cosens, Sue Jackson, Kelsey Leonard, and Daniel McCool. "Indigenous Water Justice." *Lewis and Clark Law Review* 22, no. 3 (2018): 841–921.

Roller, Duabe W. *Through the Pillars of Herakles: Greco-Roman Exploration of the Atlantic*. New York: Routledge, 2006.

Rosing, Otto, and Torben Bech, dirs. *Nuummioq*. Nuuk: 3900 Pictures, 2009.

Rud, Søren. *Colonialism in Greenland: Tradition, Governance, and Legacy*. Cham: Palgrave Macmillan, 2017.

Ruiz, Rafico. "Iceberg Economies." *TOPIA: Canadian Journal of Cultural Studies* 32 (2014): 179–99.

———. "Iceberg Media." *International Journal of Communication* 8 (2014): 2525–30.

Ryall, Anka, Johan Schimanski, and Hennign Howlid Wærp. "Arctic Discourses: An Introduction." In *Arctic Discourses*, edited by Anka Ryall, Johan Schimanski, and Hennign Howlid Wærp, ix–xxii. Newcastle upon Tyne: Cambridge Scholars Publishing, 2010.

Said, Edward. *Orientalism*. New York: Pantheon Books, 1978.

Salzman, James. *Drinking Water: A History*. New York: Overlook Duckworth, 2013.

Saxo Grammaticus. *The Nine Books of the Danish History of Saxo Grammaticus*. Vol. 1. Translated by Oliver Elton. London: Norrœna Society, 1906.

Scharf, Michael P. *Customary International Law in Times of Fundamental Change: Recognizing Grotian Moments*. Cambridge: Cambridge University Press, 2013.

Schumann, Karl. "Hobbes's Concept of History." In *Hobbes and History*, edited by G. A. J. Rogers and Tom Sorell, 3–24. Abingdon, UK: Routledge, 2000.

Segerfeldt, Fredrik. *Water for Sale: How Business and the Market Can Resolve the World's Water Crisis*. Washington, DC: Cato Institute, 2005.

Severin, Tim. *The Brendan Voyage: Sailing to America in a Leather Boat to Prove the Legend of the Irish Sailor Saints*. New York: Random House, 2000.

Shiva, Vandana. *Water Wars: Privatization, Pollution, and Profit*. Berkeley, CA: North Atlantic Books, 2016.

Singh, Anamika. "12% of Fiji's Population Does Not Have Access to Clean and Safe Drinking Water—WAF." *fijivillage*, July 15, 2018. https://www.fijivillage.com/news/12-of-Fijis-population -does-not-have-access-to-clean-and-safe-drinking-water ---WAF-9k5s2r.

Singh, Maanvi. "Drought-Hit California Moves to Halt Nestlé from Taking Millions of Gallons of Water." *Guardian*, April 27, 2021. https://www.theguardian.com/us-news/2021/apr/27/ california-nestle-water-san-bernardino-forest-drought.

Smith, Edward H. "Oceanographer's Reports." In *International Ice Observation and Ice Patrol Service in the North Atlantic Ocean*. Washington, DC: Government Printing Office, 1924.

Smith, Kenneth L., Jr., Bruce H. Robison, John J. Helly, Ronald S. Kaufmann, Henry A. Ruhl, Timothy J. Shaw, Benjamin S. Twining, and Maria Vernet. "Free-Drifting Icebergs: Hot Spots of Chemical and Biological Enrichment in the Weddell Sea." *Science* 317, no. 5837 (2007): 478–82.

Smith, Kenneth L., Jr., A.D. Sherman, T. J. Shaw, and J. Sprintall. "Icebergs as Unique Lagrangian Ecosystems in Polar Seas." *Annual Review of Marine Science* 5 (2013): 269–87.

Sonne, Birgitte. *Worldviews of the Greenlanders: An Inuit Arctic Perspective*. Fairbanks: University of Alaska Press, 2017.

Sørensen, Axel Kjær. *Denmark-Greenland in the Twentieth Century*. Copenhagen: Commission for Scientific Research in Greenland, 2006.

Sørensen, Camilla T. N. "Chinese Investments in Greenland: Promises and Risks as Seen from Nuuk, Copenhagen and Beijing." In *Greenland and the International Politics of a Changing Arctic: Postcolonial Paradiplomacy between High and Low Politics*, edited by Kristian Søby Kristensen and Jon Rahbek-Clemmensen, 83–97. Abingdon, UK: Routledge, 2018.

Staalesen, Atle. "Rosneft Moves 1 Million Ton Big Iceberg." *Barents Observer*, October 11, 2016. https://thebarentsobserver.com/en /arctic/2016/10/rosneft-moves-1-million-ton-big-iceberg.

Stone, Gregory S. *Ice Island: Expedition to Antarctica's Largest Iceberg*. Boston: New England Aquarium Press, 2003.

Straumann, Benjamin. *Roman Law in the State of Nature: The Classical Foundations of Hugo Grotius' Natural Law*. Translated by Belinda Cooper. Cambridge: Cambridge University Press, 2015.

Szwedo, Piotr. *Cross-Border Water Trade: Legal and Interdisciplinary Perspectives*. Leiden: Brill Nijhoff, 2019.

Taverniers, Pierre. "Weather Variability and Changing Sea Ice Use in Qeqertag, West Greenland, 1987–2008." In *SIKU: Knowing Our Ice: Documenting Inuit Sea Ice Knowledge and Use,* edited by Igor Krupnik, Claudio Aporta, Shari Gearheard, Gita J. Laidler, and Lene Kielsen Holm, 31–41. Dordrecht: Springer, 2010.

Thiede, Jørn, Catherine Jessen, Paul Knutz, Antoon Kuijpers, Naja Mikkelsen, Niels Nørgaard-Pedersen, and Robert F. Spielhagen. "Millions of Years of Greenland Ice Sheet History Recorded in Ocean Sediments." *Polarforschung* 80, no. 3 (2011): 141–59.

Thirlway, Hugh. *The Sources of International Law.* Oxford: Oxford University Press, 2014.

Thomas, David N., ed. *Arctic Ecology.* Hoboken, NJ: Wiley, 2021.

Thucydides. *History of the Peloponnesian War.* Translated by Rex Warner. New York: Penguin, 1972.

"Titanic" Disaster: Hearings before a Subcommittee of the Committee on Commerce. Washington, DC: Government Printing Office, 1912.

"Titanic Sinks Four Hours after Hitting Iceberg." *New York Times,* April 16, 1912.

Tupper, Seth. "A Promise Unfulfilled: Water Pipeline Stops Short for Sioux Reservation." *Guardian,* May 23, 2019. https://www .theguardian.com/us-news/2019/may/23/a-promise-unfulfilled -water-pipeline-stops-short-for-sioux-reservation.

Verman, Jenn Thornhill. *Cod Collapse: The Rise and Fall of Newfoundland's Saltwater Cowboys.* Halifax: Nimbus Publishing Limited, 2019.

Verne, Jules. *Twenty Thousand Leagues under the Sea.* Translated by F. P. Walter. London: MacMillan Collector's Library, 2010.

Vidal, John. "Water Privatisation: A Worldwide Failure?" *Guardian,* January 30, 2015. https://www.theguardian.com/global -development/2015/jan/30/water-privatisation-worldwide -failure-lagos-world-bank.

Wade, Wyn Craig. *The* Titanic: *End of a Dream.* Revised ed. New York, Penguin: 1992.

Wadhams, Peter. *Ice in the Ocean.* London: CRC Press, 2000.

Washington Times. September 3, 1914.

"Water Is 'Catalyst' for Cooperation, Not Conflict, UN Chief Tells Security Council." UN News, June 6, 2017. https://news.un.org/ en/story/2017/06/558922-water-catalyst-cooperation-not -conflict-un-chief-tells-security-council.

Wester, Julia, Kiara R. Timpano, Demet Çek, and Kenneth Broad. "The Psychology of Recycled Water: Factors Predicting Disgust and Willingness to Use." *Water Resources Research* 52, no. 4 (2016): 3212–26.

Whiteford, Linda M., Maryann Cairns, Rebecca Zarger, and Gina Larsen. "Water, Environment, and Health: The Political Ecology of Water." In *A Companion to the Anthropology of Environmental Health*, edited by Merrill Singer, 217–35. Chichester, UK: Wiley Blackwell, 2016.

Wilson, Page. "An Arctic 'Cold Rush'? Understanding Greenland's (In)dependence Question." *Polar Record* 43, no. 5 (September 2017): 512–19.

Winter, Caroline. "Towing an Iceberg: One Captain's Plan to Bring Drinking Water to 4 Million People." *Bloomberg*, June 5, 2019. https://www.bloomberg.com/news/features/2019-06-06/ towing-an-iceberg-one-captain-s-plan-to-bring-drinking -water-to-4-million-people.

World Health Organization. "Drinking-Water Key Facts." March 21, 2022. https://www.who.int/news-room/fact-sheets/detail /drinking-water.

NOTES

p. vii   **"The First International Conference and Workshop on Iceberg Utilization . . . and Other Applications."** A. A. Husseiny, ed. *Iceberg Utilization: Proceedings of the First International Conference and Workshop on Iceberg Utilization for Fresh Water Production, Weather Modification and Other Applications held at Iowa State University, Ames, Iowa, USA, October 2–6, 1977.* (New York: Pergamon Press, 1978).

p. viii   **Iceberg Transport International Ltd.** Douglas E. Kneeland, "An Alaskan Iceberg Upstages a Saudi Prince at Conference in Iowa," *New York Times*, October 7, 1977.

p. ix   **"who have never seen an iceberg before."** Iowa State University Yearbook, *The Bomb* (Ames: Bomb Publication Board, 1977), 264.

p. ix   **"beyond anything within your experience."** Kneeland, "An Alaskan Iceberg."

p. x   **"a towed berg in situ."** *The Bomb*, 265.

p. x   **snaked into the harbor.** The event is recollected in "Just the Tip of the Iceberg . . . ," *Sydney Morning Herald*, March 29, 2003, https://www.smh.com.au/national/just-tip-of-the -iceberg-20030329-gdgiee.html.

p. 2   **waterworks would be shut down.** Gisela Kaiser, *Parched: The Cape Town Drought Story* (Cham: Springer International Publishing, 2021). See also Norimitsu Onishi and Somini Sengupta, "In South Africa, Facing 'Day Zero' With No Water," *New York Times*, January 30, 2018.

p. 4   **face regular water shortages.** *2030 Water Resources Group Annual Report 2021*, 10-11, https://2030wrg.org/wp -content/uploads/2021/12/WRG-Annual-Report_2021 _RV_Sprds_2_10_22.pdf.

p. 4   **extremely high water vulnerability.** Reimagining WASH: Water Security for All," UNICEF, March 2021, available at: https://www.unicef.org/media/95241/file/water-security -for-all.pdf

p. 4   **locked away in ice caps and glaciers.** Kristi Denise Caravella and Jocilyn Danise Martinez, "Water Quality and Quantity: Globalization," in *Managing Water Resources and Hydrological Systems*, eds. Brian D. Fath and Sven Erok Jorgensen, 265–77 (Boca Raton: CRC Press, 2021), 266.

p. 4   **two hundred thousand tons.** Two hundred thousand tons of water is about equal to 48.5 million gallons of water. If 1 gallon of water weighs 8.24 pounds and there are 2,000 pounds in a ton. If the average person drinks eight cups of water each day, they consume 0.5 gallons per day and 182.5 gallons per year. Using these figures, 250,000 people drink 45.6 million gallons per year, or about the same amount of water consumed in an average iceberg floating off Canada's East Coast.

p. 4   **icebergs are calved each year.** Vivien Gornitz, *Vanishing Ice: Glaciers, Ice Sheets, and Rising Seas* (New York: Columbia University Press, 2019), 42.

p. 5   **a "Cold Rush" for icebergs could soon erupt.** The term *Cold Rush* has been used by many people before me, including, among others: Paddy Allen and Jenny Ridley, "The 'Cold Rush': Industrialisation in the Arctic," *Guardian*, July 5, 2011; Page Wilson, "An Arctic 'Cold Rush'? Understanding Greenland's (In)dependence Question," *Polar Record* 43, no. 5 (September 2017): 512–19; Martin Breum, *Cold Rush: The Astonishing True Story of the New Quest for the Polar North* (Montreal: McGill-Queen's University Press, 2018).

p. 5   **settlement in North America outside of Greenland.** Gordon Campbell, *Norse America: The Story of a Founding Myth* (Oxford: Oxford University Press, 2021). Icebergs play a surprisingly small role in most Viking sagas. A delightful exception includes the *Saga of Bard the Snowfell God*, which relates how Helga—daughter of a half troll, half giant—is shoved onto an iceberg and drafts from Iceland to Greenland. See Ralph O'Connor and Anne O'Connor, *Icelandic Histories & Romances*, 2nd ed. (Stroud: Tempus, 2006), 193–94. For more, see Eleanor Rosamund Barraclough, *Beyond the Northlands: Viking Voyages and the Old Norse Sagas* (Oxford: Oxford University Press, 2016).

p. 6   *Inferno.* For my favorite translation, see *The Divine Comedy of Dante Alighieri* Inferno, trans. Allen Mandelbaum (New York: Bantam Books, 2004).

p. 8   **mariners who sought to conquer the world.** The name has been attributed to both the navigator Bartolomeu Dias and John II, King of Portugal. Malcolm Jack notes that, despite the optimistic name, the Portuguese maintained a negative feeling about the land. See Malcom Jack, *To the Fairest Cape: European Encounters in the Cape of Good Hope* (Lewisburg, PA: Bucknell University Press, 2019), 31.

p. 9   **trade in cinnamon, cloves, and pepper.** Crowley Roger, *Conquerors: How Portugal Forged the First Global Empire* (New York: Random House, 2015).

p. 9   **took until Judgment Day.** For a good overview of the myth, see Michael Goss, *Lost at Sea: Ghost Ships and Other Mysteries* (New York: Prometheus Books, 1994), esp. 34–58.

p. 10   **the sea journey from Europe to Asia.** Hans Friedrich Heese, "Cape of Good Hope? Meeting Place of Unwilling Migrants from Africa, Asia, and Indigenous People," *Insights of Anthropology* 4, no. 1 (2020), 268–79.

p. 10   **connected far-flung places.** James C. Armstrong and Nigel A. Worden, "The Slaves, 1652–1834," in *The Shaping of South Africa Society, 1652–1840*, ed. Richard Elphick and Hermann Giliomee (Middletown, CT: Wesleyan University Press, 1979), 109–83, esp. 143–61.

p. 14   **the International Ice Patrol (IIP).** The International Ice Patrol has since relocated to Washington, D.C.

p. 14   **1960s report put it.** Werner Bamberger, "Icebergs Defy New Coast Guard Assaults," *New York Times*, July 17, 1960.

p. 14   **"be repeated in the future."** Thucydides, *History of the Peloponnesian War*, trans. Rex Warner (New York: Penguin, 1972), 48.

p. 14   **"providently towards the future."** Quoted in Karl Schumann, "Hobbes's Concept of History," in *Hobbes and History*, ed. G. A. J. Rogers and Tom Sorell (Abingdon, UK: Routledge, 2000), 14.

p. 14   **for food, energy, and natural resources.** For a glimpse into the future, see Mary Hoff, "Envision 2050: The Future of Oceans," Ensia, May 28, 2015.

p. 17   **crushed by a rolling mass.** Writing in the *Gesta Danorum* around 1200, Saxo describes a calving iceberg on the "squalid" island west of Norway, likely Iceland. He recounts the "extraordinary clamour" and how the "mind stands dazed in wonder" by the magnitude of the iceberg's break from the glacier (87). The men standing too close, Saxo explains, were crushed and pulled into the sea as the ice rolled, eventually reemerging only to be "picked up dead." See Saxo Grammaticus, *The Nine Books of the Danish History of Saxo Grammaticus*, vol. 1, trans. Oliver Elton (London: Norrœna Society, 1906), 87.

p. 17   **5.0 magnitude earthquake.** T. C. Bartholomaus, C. F. Larsen, S. O'Neel, and M. E. West, "Calving Seismicity from Iceberg–Sea Surface Interactions," *Journal of Geophysical Research* (December 22, 2012), doi:10.1029/ 2012JF002513.

p. 18   **the *Guiding Star.*** Details are recorded in "The Guiding Star," *Sydney Morning Herald*, September 10, 1855.

p. 18   **toward which the *Guiding Star* was cruising.** The captain of the *George Marshall* described the immense quantities of icebergs in the sea and the "unbroken line of icebergs" three miles long, which stretched across the intended path of the *Guiding Star*, in the Shipping Intelligence section of the Melbourne-based newspaper the *Argus*, April 7, 1855.

p. 19   **"collision with the ice."** Quoted in Kenneth N. Mival, "Re-Assessing the Disappearance of the Clipper 'Guiding Star,'" *Great Circle* 37, no. 2 (2015): 17.

p. 19   **from Liverpool for New York.** For more details about the fascinating life of Asa Eldridge, see Vincent Miles, *The Lost Hero of Cape Cod: Captain Asa Eldridge and the Maritime Trade That Shaped America* (Yarmouth Port, MA: Historical Society of Old Yarmouth, 2015).

p. 19   ***please get it published.*** Cited in Wyn Craig Wade, *The Titanic: End of a Dream*, revised ed. (New York, Penguin: 1992), 32.

p. 19   **the story of the "unsinkable" ship *Titan.*** Published in 1898 as *Futility*, the text was revised and republished as *The Wreck of the Titan* in 1912.

p. 20   **foreshadows on page five.** Morgan Robertson, *Futility* (New York: M. F. Mansfield, 1898), 5.

p. 20   **screams of women and children.** Ibid., 54.

p. 20   **gifted with precognition.** Michelle McQuigge, "Author's Fictional Titan Hit Iceberg Years before the *Titanic* Sank," *Canadian Press*, April 12, 2012.

p. 20   **ocean liner without lifeboats.** The novel was published in
         an English translation the same year.

p. 20   **"Isn't man's courage utter madness?"** Gerhart
         Hauptmann, *Atlantis*, trans. Adele and Thomas Seltzer
         (New York: B. W. Huebsch, 1912), 95.

p. 20   **"in the fog almost submerged?"** Ibid., 95.

p. 20   **"last through the ages."** "Hauptmann's Prize," *New York
         Times*, November 16, 1912.

p. 21   **2,240 passengers and crew on board.** John Wilson Foster
         has written an engaging account of the *Titanic* and situated
         the ship in its broader cultural context. See John Wilson
         Foster, Titanic: *Culture and Calamity* (Vancouver: Belcouver
         Press, 2016).

p. 21   **"Iceberg! Right ahead!" he warned.** *"Titanic" Disaster:
         Hearings before a Subcommittee of the Committee on Commerce*
         (Washington, DC: Government Printing Office, 1912),
         317, 450.

p. 21   **"narrow shave."** Ibid., 321.

p. 22   **a logical explanation.** "Titanic Sinks Four Hours after
         Hitting Iceberg," *New York Times*, April 16, 1912.

p. 22   **ignored on the bridge.** Senan Molony reconstructs the
         various causes that led to the sinking of the *Titanic*, telling
         a story far more complex than the version I learned growing
         up. See Senan Molony, Titanic: *Why She Collided, Why She
         Sank, Why She Should Never Have Sailed* (Cork: Mercier
         Press, 2019).

p. 23   **"a service of ice patrol."** International Convention of the
         Safety of Life at Sea, translation, chapter 3, article 6, in
         *International Conference on Safety of Life at Sea: Messages
         from the President* (Washington, DC: Government Printing
         Office, 1914), 9. "Ocean" and create "a service for the
         study and observation of ice conditions and a service of ice
         patrol."

p. 23   **powder keg at the time.** My favorite account of Europe
         at the time is Christopher Clark's spellbinding *The
         Sleepwalkers: How Europe Went to War in 1914* (New York:
         HarperCollins, 2013).

p. 24   **eliminate the threat altogether.** For more on the
         commander, see Robert Irwin Johnson, *Guardians of the Sea:
         History of the United States Coast Guard, 1915 to the Present*
         (Annapolis, MD: Naval Institute Press, 1987, esp. 116–25.

p. 26   **"more than one day, possibly two."** Edward H.
         Smith, "Oceanographer's Reports," in *International Ice
         Observation and Ice Patrol Service in the North Atlantic Ocean*
         (Washington, DC: Government Printing Office, 1924), 83.

p. 26   **"destructive man-made means."** Werner Bamberger,
         "International Ice Patrol Ends; Coast Guard Spend Busy
         Season; Attempts to Destroy Bergs by Bombing Not
         Successful—Data Gathered to Improve Detection by
         Radar," *New York Times*, July 26, 1959.

p. 27   **approaching an iceberg.** Werner Bamberger, "Icebergs
         Defy New Coast Guard Assaults," *New York Times*, July 17,
         1960.

p. 28   **"the vanity of man."** "High Seas: Little *Titanic*," *Time*,
         February 9, 1959.

p. 29   **cost a mind-boggling sum.** In my attempt to provide an
         exact number, I went down the complicated rabbit hole of
         sea transportation costs. The cost of a delay is influenced by
         many variables. The work of David Hummels is considered
         by many to be field-leading. See David Hummels,
         "Transportation Costs and International Trade in Second
         Era of Globalization," *Journal of Economic Perspectives* 21,
         no. 3 (2007): 131–54.

p. 37   **boulders around the world.** Charles Lyell, *Principles of
         Geology*, vol. 1 (London: John Murray, 1830), 117–20.

p. 37   **the dispersion of species.** In *The Origin of Species*, Charles
         Darwin writes: "As icebergs are known to sometimes
         be loaded with earth and stones, and have even carried
         brushwood, bones, and the nest of a land-bird, I can
         hardly doubt that they must occasionally have transported
         seeds from one part to another of the arctic and Antarctic
         regions, as suggested by Lyell. . . . Hence we may safely
         infer that icebergs formerly landed their rocky burthens
         on the shores of these mid-ocean islands, and it is at least
         possible that they may have brought thither the seeds of
         northern plants. . . . Moreover, icebergs formerly brought
         boulders to its western shores, and they may have formerly
         transported foxes, as so frequently now happens in the
         arctic regions." Charles Darwin, *The Origin of Species*
         (Oxford: Oxford University Press, 1998), 293, 294, 318.
         See also Peter J. Bowler, "Geographical Distribution in the
         Origin of Species," in *The Cambridge Companion to the "Origin*

of Species," ed. Michael Ruse and Robert J. Richards, 153–72 (Cambridge: Cambridge University Press, 2009), 161.

p. 37   **woolly mammoths frozen inside icebergs.** See, for example, William Rhind, *Elements of Geology and Physical Geography* (Edinburgh: Fraser & Co. J. Anderson, 1837), 52. Lyell discusses the phenomenon as well, in *Principles of Geology*, vol. 1, 97–98.

p. 37   **still gripping his spear and shield.** "Entombed in an Iceberg for Centuries," *Cincinnati Enquirer*, March 26, 1922.

p. 37   **eat snow in urban areas.** Ariya A. Parisa et al., "Role of Snow and Cold Environment in the Fate and Effects of Nanoparticles and Select Organic Pollutants from Gasoline Engine Exhaust," *Environmental Sciences: Processes and Impacts* 18, no. 2 (2016): 153–296. Or, for a less scientific take, see Sarah Kaplan, "Snow Is Beautiful. Please Don't Eat It," *Washington Post*, January 22, 2016, https://www .washingtonpost.com/news/morning-mix/wp/2016/01/22/ snow-is-beautiful-please-dont-eat-it.

p. 37   **icebergs as having a life cycle.** One of the most helpful books I consulted to learn about icebergs was Grant R. Biggs, *Icebergs: Their Science and Links to Global Change* (Cambridge: Cambridge University Press, 2015).

p. 38   **"singing" icebergs.** Christian Müller et al., "Singing Icebergs," *Science* 310, no. 5727 (2005): 1299.

p. 39   **just 320 nautical miles away.** For a nice account of the glacier, see Jon Gertner, *The Ice at the End of the World: An Epic Journey into Greenland's Buried Past and Our Perilous Future* (New York: Random House, 2020), 259.

p. 39   **went belly-up.** Jenn Thornhill Verman offers a beautifully human look at the situation in *Cod Collapse: The Rise and Fall of Newfoundland's Saltwater Cowboys* (Halifax: Nimbus Publishing Limited, 2019). For a history more focused on the fish, see Michael Harris, *Lament for an Ocean: The Collapse of the Atlantic Cod Fishery* (Toronto: McCelland & Stewart, 2013).

p. 45   **"revolving on its axis, as a spit?"** Herman Melville, *Redburn* (New York: Random House, 2002), 110.

p. 45   **"the motion of the ship."** Ibid., 118.

p. 45   **Perfect Storm of 1991.** Enthrallingly memorialized in Sebastian Junger, *The Perfect Storm: A True Story of Men against the Sea* (New York: W. W. Norton, 1997).

p. 46   **a cafeteria, gym, and swimming pool.** Shamseer
          Mambra, "5 Biggest Oil Platforms in the World," Marine
          Insight, March 9, 2022, https://www.marineinsight.com/
          offshore/5-biggest-oil-platforms-in-the-world.

p. 47   **standing 8 meters above the waterline.** "Iceberg in Near-
          Miss with Floating Oil Platform off Newfoundland,"
          Hamilton Spectator, March 30, 2017, https://www.thespec.
          com/news/canada/2017/03/30/iceberg-in-near-miss-with-
          floating-oil-platform-off
          -newfoundland.html.

p. 49   **abort their mission and turn around.** Jim Gomez, "China
          Coast Guard Uses Water Cannon against Philippine
          Boats," Associated Press, November 18, 2021, https://
          apnews.com/article/china-united-states-philippines
          -manila-south-china-sea-9fe8af0a7ae2386e058e99a7c
          5c33243.

p. 50   **lasso a one-million-ton iceberg.** Atle Staalesen, "Rosneft
          Moves 1 Million Ton Big Iceberg," *Barents Observer,*
          October 11, 2016, https://thebarentsobserver.com/en/
          arctic/2016/10/rosneft-moves-1-million-ton-big-iceberg.

p. 50   **"economically driven."** Chris O'Neill-Yates, "Investigation
          Shows Husky's Decision Not to Move Out of Iceberg's
          Path in Close Call 'Economically Driven,'" *CBC News,*
          February 26, 2018, https://www.cbc.ca/news/canada/
          newfoundland-labrador/husky-iceberg-suspension-searose
          -fpso-suspension-cnlopb-offshore-1.4551362.

p. 52   **the heist had been carefully planned.** Leyland Cecco,
          "Spirited Away: Canadian Thieves Steal More Than
          $9,000 in Iceberg Water," *Guardian,* February 14, 2019,
          https://www.theguardian.com/world/2019/feb/14/
          canada-iceberg-water-theft-vodka.

p. 53   **"deserves epicurean attention."** The Fine Water Society
          was founded in 2008. More can be found on their website:
          https://finewaters.com.

p. 54   **no visa required.** The act, Lov om Svalbard, was passed by
          the Norwegian Parliament on July 17, 1925, and entered
          into force on August 14, 1925.

p. 54   **"chasing off polar bears."** The governor of Svalbard's
          *Guidelines for Firearms and Protection and Scaring Devices
          against Polar Bears,* adopted and entered into force on
          20.07.2021 pursuant to Act No. 7 of April 20, 2018, offers

a fascinating look into the daily reality of people who live on the Arctic island.

p. 56 **not the only possible point of view.** The first to use this term may well be Anka Ryall, Johan Schimanski, and Hennign Howlid Wærp, "Arctic Discourses: An Introduction," in *Arctic Discourses*, ed. Anka Ryall, Johan Schimanski, and Hennign Howlid Wærp (Newcastle upon Tyne: Cambridge Scholars Publishing, 2010), x. The term is borrowed from Edward Said, *Orientalism* (New York: Pantheon Books, 1978).

p. 56 **a frozen sea, the midnight sun, and the aurora.** Although the Greek explorer's text, *On the Ocean*, has been lost, a number of authors cite Pytheas and his travels. See Duabe W. Roller, *Through the Pillars of Herakles: Greco-Roman Exploration of the Atlantic* (New York: Routledge, 2006), 57–91. For a compelling history of the fabled land of Thule described by Pytheas, see Joanna Kavenna, *The Ice Museum: In Search of the Lost Land of Thule* (New York: Penguin, 2007).

p. 56 **the Garden of Eden.** *Navigatio Sancti Brendani Abbatis*, trans. John O'Meara and Jonathan Wooding, in *The Voyage of Saint Brendan: Representative Versions of the Legend in English Translation*, ed. W. R. J. Barron and Glyn S. Burgess (Exeter: University of Exeter Press, 2002). The late British explorer and author Tim Severin brilliantly recreated St. Brendan's journey. He built a replica of the currach and successfully crossed the Atlantic from Ireland to Canada from May 1976 to June 1977. His voyage is recorded in Tim Severin, *The Brendan Voyage: Sailing to America in a Leather Boat to Prove the Legend of the Irish Sailor Saints* (New York: Random House, 2000).

p. 57 **the depths of the ocean.** For an erudite analysis of the "crystal pillar" see Jude S. Mackley, *The Legend of St. Brendan: A Comparative Study of the Latin and Anglo-Norman Versions* (Leiden: Brill, 2008), 177–86.

p. 58 **"countless minarets and mosques."** Jules Verne, *Twenty Thousand Leagues under the Sea*, trans. F. P. Walter (London: MacMillan Collector's Library, 2010), 405.

p. 58 **"veins running inside the ice."** Ibid., 438.

p. 58 **"the unsuspecting blood-vessels."** Quoted in Jennifer Raab, "'Precisely These Objects': Frederic Church and the Culture of Detail," *Art Bulletin* 95, no. 4 (2013): 588.

p. 58 **the mysteries of the Arctic.** For a persuasive analysis of the artwork and a history of its creation, see: Gerald L. Carr, *Frederic Edwin Church: The Icebergs* (Dallas: Dallas Museum of Fine Arts, 1980).

p. 59 **sell its fantasy fur looks.** Laird Borrelli-Persson, "The Big Chill: Revisiting Chanel's Fall 2010 Iceberg Show," *Vogue*, December 27, 2018, https://www.vogue.com/article/karl -lagerfeld-imported-an-iceberg-from-sweden-for-his-fall -2010-chanel-show.

p. 65 **specialization in food.** Mascha has written an educational book about water and describes icebergs as a source: Michael Mascha, *Fine Waters: A Connoisseur's Guide to the World's Most Distinctive Bottled Waters* (Philadelphia: Quirk Books, 2006), 29.

p. 66 **water sommelier, author, and media personality.** If you are interested in learning more about water, I heartily recommend browsing Martin's website: www.martin -riese.com.

p. 69 **sustainability is buzzy.** Andrea Mull, "How Fancy Water Bottles Became a 21st-Century Status Symbol," *Atlantic*, February 12, 2019, https://www.theatlantic.com/health/ archive/2019/02/luxury-water-bottles/582595.

p. 69 **mainstream advertising.** For a small taste of the fascinating business of luxury goods, see Martin Fassnacht, Philipp Nikolaus Kluge, and Henning Mohr, "Pricing Luxury Brands: Specificities, Conceptualization and Performance Impact," *Marketing: ZFP—Journal of Research and Management* 35, no. 2 (2013): 104–17.

p. 73 **belongs to everyone as a public resource.** There is a wealth of terrific scholarship on the issues related to the ownership and use of water. James Salzman's history provides a nice overview of the various ideas. See James Salzman, *Drinking Water: A History* (New York: Overlook Duckworth, 2013), esp. 46–71, as does Emanuele Fantini, "An Introduction to the Human Right to Water: Law, Politics, and Beyond," *WIREs Water* 7, no. 2 (2020), https://wires.onlinelibrary .wiley.com/doi/10.1002/wat2.1405.

p. 73 **cuts into the bottom line.** Unless you have spent a great deal of time contemplating water, Maude Barlow's work will fundamentally change the way you think about it. See Maude Barlow, *Blue Covenant: The Global Water Crisis and* *the Coming Battle for the Right to Water* (New York: The New Press, 2007).

p. 73   **to market the purity of its product.** Maanvi Singh, "Drought-Hit California Moves to Halt Nestlé from Taking Millions of Gallons of Water," *Guardian*, April 27, 2021, https://www.theguardian.com/us-news/2021/apr/27 /california-nestle-water-san-bernardino-forest-drought; Ameilia Gentleman, "Coca-Cola Urged to Close an Indian Plant to Save Water," *New York Times*, January 16, 2008, https://www.nytimes.com/2008/01/16/business/16coke .html; Charles Fishman, "Message in a Bottle," *Fast Company*, July 1, 2007, https://www.fastcompany.com/ 59971/message-bottle.

p. 73   **the needs of the poorest people.** Vandana Shiva offers an overview. See Vandana Shiva, *Water Wars: Privatization, Pollution, and Profit* (Berkeley, CA: North Atlantic Books, 2016), 19–38. Water privatization in the Kathmandu Valley offers an interesting case study. See Aditi Raina, Yogendra Gurung, and Bhim Suwal, "Equity Impacts of Informal Private Water Markets: Case of Kathmandu Valley," *Water Policy* 22 (2020): 189–204. For a look at privatization in the United States, see Elizabeth Douglas, "Towns Sell Their Public Water Systems—and Come to Regret It," *Washington Post*, July 8, 2017, https://www.washingtonpost .com/national/health-science/towns-sell-their-public-water -systems--and-come-to-regret-it/2017/07/07/6ec5b8d6 -4bc6-11e7-bc1b-fddbd8359dee_story.html. For a global take, see John Vidal, "Water Privatisation: A Worldwide Failure?" *Guardian*, January 30, 2015, https://www.the guardian.com/global-development/2015/jan/30/water -privatisation-worldwide-failure-lagos-world-bank.

p. 73   **a reliable source of clean freshwater.** According to the Water Authority of Fiji. Anamika Singh, "12% of Fiji's Population Does Not Have Access to Clean and Safe Drinking Water— WAF" *fijivillage*, July 15, 2018, https://www.fijivillage.com/ news/12-of-Fijis-population-does-not-have-access-to-clean -and-safe-drinking-water---WAF-9k5s2r.

p. 75   **"glands of the throat."** Entry from January 9, 1773, Georg Forster, *A Voyage Round the World*, vol. 1, ed. Nicholas Thomas and Oliver Berghof (Honolulu: University of Hawai'i Press, 2000), 71.

p. 75   **jagged coast of Newfoundland.** My favorite biography of
            James Cook is Frank McLynn, *Captain Cook: Master of the
            Seas* (New Haven, CT: Yale University Press, 2011).

p. 76   **"a little tedious."** James Cook, *The Journals of Captain Cook*,
            ed. Philip Edwards (London: Penguin 2003), 156.

p. 76   **"we had on board."** Entry from January 9, 1773, cited in *A
            Voyage Round the World*, 71.

p. 76   **"what he most wanted."** John Pringle, "A Discourse upon
            Some Late Improvements of the Means for Preserving the
            Health of Mariners," delivered at the Anniversary Meeting
            of the Royal Society, November 30, 1776, in *A Voyage
            towards the South Pole and Round the World*, vol. 2, by James
            Cook (London: Strahan and Cadell, 1777), 393.

p. 78   **$2 billion in damage.** "Costa Concordia Capsizing
            Costs over $2 Billion for Owners," Reuters, July 2, 2014,
            https://www.reuters.com/article/italy-concordia-costs
            /costa-concordia-capsizing-costs-over-2-billion-for-owners
            -idUSL6N0PH0EO20140706.

p. 78   **to get the job done.** For an account of the undertaking, see
            "The Brains behind the Costa Concordia Salvage," *Local*,
            September 17, 2013, https://www.thelocal.it/20130917/
            the-mastermind-behind-the-concordia-salvage-operation.

p. 78   **"born optimist."** Quoted in the laudation for the *Deutscher
            Meerpreis* at https://www.deutscher-meerespreis.de/2015
            _en.html.

p. 81   **Project Habakkuk.** Quoted from Andrew Roberts, *Churchill:
            Walking with Destiny* (New York: Penguin, 2018), 169.

p. 81   **planes could land and shelter.** For a history of the
            extraordinary story of Geoffrey Pyke, see Henry Hemming,
            The Ingenious Mr. Pyke: Inventor, Fugitive, Spy (New York:
            PublicAffairs, 2015).

p. 82   **"need to be discussed."** Quoted in Mariana Gosnell,
            *Ice: The Nature, the History, and the Uses of an Astonishing
            Substance* (New York: Knopf, 2011).

p. 82   **flopped or were part of a stunt.** Peter Wadhams, *Ice in the
            Ocean* (London: CRC Press, 2000), 267.

p. 82   **dismissed the idea as harebrained.** "A Genius in Bedford,"
            Scientific American, August 22, 1863.

p. 83   **made fun of the people who wrote in.** For more on the
            incident and the history of iceberg towing, see Alexis C.
            Madrigal, "The Many Failures and Few Successes of Zany

Iceberg Towing Schemes," *Atlantic*, August 10, 2011, https://www.theatlantic.com/technology/archive/2011/08/ the-many-failures-and-few-successes-of-zany-iceberg -towing-schemes/243364.

p. 83   **Boston, New York, Baltimore, and Philadelphia.** *Washington Times*, September 3, 1914.

p. 83   **created chaos wherever they went.** *Boston Globe*, September 3, 1914.

p. 84   **promotional videos visualizing the plan.** My favorite such video can be found at https://www.youtube.com/ watch?v=CdC3HPNjc94.

p. 85   **"the iceberg would have to be big."** Quoted in Caroline Winter, "Towing an Iceberg: One Captain's Plan to Bring Drinking Water to 4 Million People," *Bloomberg*, June 5, 2019, https://www.bloomberg.com/news/features/2019 -06-06/towing-an-iceberg-one-captain-s-plan-to-bring -drinking-water-to-4-million-people.

p. 85   **the same size as Jamaica.** Gregory S. Stone, with the help of photographer Wes Skiles, captured breathtaking photographs of the iceberg in Gregory S. Stone, *Ice Island: Expedition to Antarctica's Largest Iceberg* (Boston: New England Aquarium Press, 2003).

p. 86   **only lose 38 percent of its mass.** Reported by Mark Brown, "How to Tow a Building-Sized Iceberg," *Wired*, August 10, 2011, https://www.wired.com/2011/08/iceberg-towing -drinking-water.

p. 88   **headed by Timm Schwarzer and Heiner Schwer.** https:// www.polewater.com.

p. 92   **"not believe, even if you were told."** Habakkuk 1:5, *The New Oxford Annotated Bible with Apocrypha* (Oxford: Oxford University Press, 2018), 1342.

p. 92   **"composed of, if you know?"** *"Titanic" Disaster: Hearings before a Subcommittee of the Committee on Commerce*, 380.

p. 93   **may have been wanting.** Daniel Allen Butler, *The Other Side of the Night: The Carpathia, the Californian, and the Night the Titanic Was Lost* (Philadelphia: Casemate, 2009), 143.

p. 96   **known as the Spice Islands.** Herman Melville, *Moby-Dick; Or, The Whale* (New York: Penguin, 2003), 12.

p. 96   **Miocene Epoch.** Andrew Carter et al., "Widespread Antarctic Glaciation during the Late Eocene," *Earth and Planetary Science Letters* 458 (2017): 49–57; Jørn Thiede et al.,

"Millions of Years of Greenland Ice Sheet History Recorded in Ocean Sediments," *Polarforschung* 80, no. 3 (2011): 141–59. For more on the history, see Roger G. Barry and Eileen A. Hall-McKim, *Polar Environments and Global Change* (Cambridge: Cambridge University Press, 2018), 21–66.

p. 97   **Earth's chemical, biological, and physical systems.** For a scholarly examination of the history of the name and concept, see Susanne Benner et al., eds., *Paul J. Crutzen and the Anthropocene: A New Epoch in Earth's History* (Cham: Springer, 2021). In my own thinking and writing, I typically follow Lewis and Maslin's later start date for the Anthropocene, 1964. See Simon L. Lewis and Mark A. Maslin, "Defining the Anthropocene," *Nature* 519 (2015): 171–80.

p. 97   **Mariana Trench.** Laura Parker, "Microplastics Have Moved into Virtually Every Crevice on Earth," *National Geographic*, August 7, 2020, https://www.nationalgeographic.com/science/article/microplastics-in-virtually-every-crevice-on-earth.

p. 97   **the past twelve thousand years.** Christopher Intagliata, "Greenland Is Melting Faster Than Any Time in Past 12,000 Years," *Scientific American*, October 3, 2020, https://www.scientificamerican.com/podcast/episode/greenland-is-melting-faster-than-any-time-in-past-12-000-years.

p. 102   **far from lifeless.** David N. Thomas, ed., *Arctic Ecology* (Hoboken, NJ: Wiley, 2021) helped me to better understand the variety of lifeforms in the Arctic and the region's connection to the rest of the planet.

p. 102   **seawater, sea ice, and sediment.** The *State of the Arctic Marine Biodiversity Report* by the Conservation of Arctic Flora and Fauna is a terrific resource for learning about life in the Arctic, https://www.arcticbiodiversity.is.

p. 102   **components of this ecosystem.** Kenneth L. Smith Jr. et al., "Free-Drifting Icebergs: Hot Spots of Chemical and Biological Enrichment in the Weddell Sea," *Science* 317, no. 5837 (2007): 478–82.

p. 103   **lasts several weeks.** Kenneth L. Smith Jr. et al., "Icebergs as Unique Lagrangian Ecosystems in Polar Seas," *Annual Review of Marine Science* 5 (2013): 276.

p. 104   **"possibly significant proportions."** Ibid., 273.

p. 104   **"organic carbon as fecal material."** Ibid., 274.

p. 108   **mink whales in the North Sea.** Fred Mogul, "Offshore Wind May Help the Planet—But Will It Hurt Whales,"

NPR, December 5, 2019, https://www.npr.org/2019/12/05/
782694371/offshore-wind-may-help-the-planet-but-will
-it-hurt-whales.

p. 110   **2.1 tons each year on average.** European Environment
Agency, *European Maritime Transport Environmental
Report* (Luxembourg: Publication Office of the European
Union, 2021), https://www.eea.europa.eu/publications/
maritime-transport.

p. 110   **Arctic summer sea ice.** Dirk Notz and Julienne Stroeve,
"Observed Arctic Sea-Ice Loss Directly Follows
Anthropogenic CO_2 Emission," *Science* 354, no. 6313
(2016): 747–50.

p. 111   **environmental friendliness.** Guy Pearse offers a
fascinating and comprehensive look at greenwashing,
investigating a variety of industries, companies, products,
and people to determine just how climate friendly they
really are. Though outdated now—hopefully businesses
have updated their environmental policies to match
their marketing—the book helps the reader identify
greenwashing practices and better evaluate businesses
before buying purportedly "green" products. See Guy
Pearse, *The Greenwash Effect: Corporate Deception, Celebrity
Environmentalists, and What Big Business Isn't Telling You
about Their Green Products and Brands* (New York: Skyhorse
Publishing, 2014).

p. 111   **the Stockholm Environment Institute explains.** Derik
Broekhoff et al., "Securing Climate Benefit: A Guide to
Using Carbon Offsets," Stockholm Environment Institute
and Greenhouse Gas Management Institute, 2019, www
.offsetguide.org.

p. 113   **South African ports every year.** Thean Potgieter, "South
Africa: The Blue Economy Experience," in *The Blue Economy
in Sub-Saharan Africa: Working for a Sustainable Future*, ed.
Donald L. Sparks (New York: Routledge, 2021), 125.

p. 113   **endangered whales fed.** Eliza Oldach et al., "Managed
and Unmanaged Whale mortality in the California
Current Ecosystem," *Marine Policy* 140 (June 2022),
https://www.sciencedirect.com/science/article/pii/
S0308597X22000860#bbib4.

p. 113   **drive global temperatures up.** Dahr Jamail's book is a
lovely meditation on ice and will inspire you to take as

much in before you lose the chance, even if he does not
focus on icebergs or ice shelves. See Dahr Jamail, *The End
of Ice: Bearing Witness and Finding Meaning in the Path of
Climate Disruption* (New York: The New Press, 2019).

p. 114   **global climate governance.** Val Plumwood, *Environmental
Culture: The Ecological Crisis of Reason* (New York:
Routledge, 2002), esp. 71–80.

p. 117   **since the Stone Age.** For a history in English, see Naja
Mikkelsen and Torsten Ingerslev, eds., *Nomination of
the Ilulissat Icefjord for Inclusion in the World Heritage
List* (Copenhagen: Geological Survey of Denmark and
Greenland, 2002), https://whc.unesco.org/uploads/
nominations/1149.pdf.

p. 117   **its environmental policies.** For record of the changing
legal status of Greenland, see Axel Kjær Sørensen,
Denmark-Greenland in the Twentieth Century (Copenhagen:
Commission for Scientific Research in Greenland, 2006).

p. 117   **the situatedness of life.** The term *situatedness* has been
used in sundry ways over the past three decades. For an
overview, see Thomas Hünefeldt and Annika Schlitte,
"Introduction: Situatedness and Place," in *Situatedness and
Place: Multidisciplinary Perspective on the Spacio-Temporal
Contingency of Human Life*, ed. Thomas Hünefeldt and
Annika Schlitte (Cham: Springer, 2018), 1–18.

p. 118   **"How is water situated?"** See, for example, Rutgerd
Boelens, Jeroen Vos, and Tom Perrault, "Introduction:
The Multiple Challenges and Layers of Water Justice
Struggles," in *Water Justice*, ed. Rutgerd Boelens, Jeroen
Vos, and Tom Perrault (Cambridge: Cambridge University
Press, 2018), 15; Linda M. Whiteford et al., "The Political
Ecology of Water," in *A Companion to the Anthropology of
Environmental Health*, ed. Merrill Singer (Chichester, UK:
Wiley Blackwell, 2016), 221.

p. 122   **"for those perishing at sea."** Hinrich Rink, *Tales and
Traditions of the Eskimo with a Sketch of their Habits, Religion,
Language and Other Peculiarities*, trans. Robert Brown
(Cambridge: Cambridge University Press, 2014), 417.

p. 122   **icebergs as a water source.** Birgitte Sonne, *Worldviews
of the Greenlanders: An Inuit Arctic Perspective* (Fairbanks:
University of Alaska Press, 2017), 405; Pierre Taverniers,
"Weather Variability and Changing Sea Ice Use in Qeqertag,

West Greenland, 1987–2008," in *SIKU: Knowing Our Ice: Documenting Inuit Sea Ice Knowledge and Use*, ed. Igor Krupnik et al. (Dordrecht: Springer, 2010), 36–37.

p. 122   **shorthand for Greenland itself.** For a brief overview of Greenlandic art and culture with great images, see Ole G. Jensen, *The Culture of Greenland in Glimpses* (Nuuk: Milik Publishing, 2007).

p. 122   **film produced entirely in Greenland.** Otto Rosing and Torben Bech, dirs., *Nuummioq* (3900 Pictures, Nuuk, 2009).

p. 122   **"stunning triumph."** James Greenberg, "Greenland's 'Nuummioq' a Stunning Triumph," Reuters, January 25, 2010, https://www.reuters.com/article/us-film-nuummioq/ greenlands-nuummioq-a-stunning-triumph-idUSTRE60 P0EA20100126.

p. 125   **other Indigenous communities.** Naja Dyrendom Graugaard, "'Without Seals, There Are No Greenlanders,' Colonial and Postcolonial Narratives of Sustainability and Inuit Seal Hunting," in *The Politics of Sustainability in the Arctic: Reconfiguring Identity, Space, and Time*, ed. Jeppe Stransbjerg and Ulrik Pram Gad (New York: Routledge, 2019), 74–93.

p. 127   **troubling history in this regard.** Few books about the history of Greenland have been written in English. Luckily, we have Rud's tome. See Søren Rud, *Colonialism in Greenland: Tradition, Governance and Legacy* (Cham: Palgrave Macmillan, 2017).

p. 128   **and violence followed.** Janne Flora has written a beautiful study of Inuit communities in Greenland and the high suicide rates that plague them. See Janne Flora, *Wandering Spirits: Loneliness and Longing in Greenland* (Chicago: University of Chicago Press, 2019).

p. 128   **most of them within twenty years.** Martin Breum, "Her er den egentlige forskel på dansk og grønlandsk syn på fremtiden," *Altinget*, January 9, 2019, https://www.altinget .dk/arktis/artikel/martin-breum-her-er-den-egentlige -forskel-paa-dansk-og-groenlandsk-syn-paa-fremtiden.

p. 128   **most powerful party in the country.** Jacob Gronholt -Pedersen, "Left-Wing Warty Wins Greenland Election, Opposes Big Mining Project," Reuters, April 7, 2021, https://www.reuters.com/article/us-greenland-election/left -wing-party-wins-greenland-election-opposes-big-mining -project-idUSKBN2BU0V1.

p. 128   **rare earth elements on the island.** Camilla T. N. Sørensen
          has written a helpful primer on the various perspectives
          of Chinese investments in Greenland. See Camilla T. N.
          Sørensen, "Chinese Investments in Greenland: Promises
          and Risks as Seen from Nuuk, Copenhagen and Beijing," in
          *Greenland and the International Politics of a Changing Arctic:
          Postcolonial Paradiplomacy between High and Low Politics*,
          ed. Kristian Søby Kristensen and Jon Rahbek-Clemmensen
          (Abingdon, UK: Routledge, 2018), 83–97.

p. 129   **to fill up buckets to take home.** Seth Tupper, "A Promise
          Unfulfilled: Water Pipeline Stops Short for Sioux
          Reservation," *Guardian*, May 23, 2019, https://www.the
          guardian.com/us-news/2019/may/23/a-promise-unfulfilled
          -water-pipeline-stops-short-for-sioux-reservation.

p. 129   **lack complete plumbing in their homes.** Laura Morales,
          "Many Native Americans Can't Get Clean Water, Report
          Finds," NPR, November 18, 2019, citing Dig Deep and US
          Water Alliance, *Closing the Water Access Gap in the United
          States: A National Action Plan*, 2019, 12, https://www
          .digdeep.org/close-the-water-gap.

p. 130   **Native Americans were excluded.** Environmental crises
          are often the result of a poor understanding of policy and
          science, and also, as Kuhn and Fleck show in the case of
          the management of the Colorado River, willful ignorance
          and the selective embrace of available information. See Eric
          Kuhn and John Fleck, *Science Be Damned: How Ignoring
          Inconvenient Science Drained the Colorado River* (Tucson:
          University of Arizona Press, 2019). Many histories of the
          Colorado River Compact make only brief reference to
          the exclusion of Native Americans, including the seminal
          work on the topic: Norris Hundley, *Water and the West:
          The Colorado River Compact and the Politics of Water in the
          American West* (Berkeley: University of California Press,
          2009), 334.

p. 130   **the basin's market value.** Lana D. Hartwig et al.,
          "Benchmarking Indigenous Water Holdings in the
          Murray-Darling Basin: A Crucial Step towards Developing
          Water Rights Targets for Australia," *Australasian Journal of
          Water Resources* 25, no. 2 (2021): 98–110.

p. 130   **reversal of the privatization.** Oscar Olivera, *Cochabamba! Water
          War in Bolivia* (Cambridge, MA: South End Press, 2004).

p. 130   **Indigenous water justice in particular.** Jason Robison et al., "Indigenous Water Justice," *Lewis and Clark Law Review* 22, no. 3 (2018): 841–921.

p. 130   **right to self-determination.** Declaration on the Rights of Indigenous Peoples, G.A. Resolution 61/295, UN Doc A /RES/61/295, September 13, 2007, Articles 25, 26(1), 32(1).

p. 131   **availability and purity of water.** *Indigenous Peoples Kyoto Water Declaration*, March 2003.

p. 132   **rich diversity of Indigenous life.** For a nice discussion of the trope, see Michael E. Harkin and David Rich Lewis, "Introduction," in *Native Americans and the Environment: Perspectives on the Ecological Indian*, ed. Michael E. Harkin and David Rich Lewis (Lincoln: University of Nebraska Press, 2007), xix–xxxiv.

p. 135   **"the property of the first taker."** Quoted in Benjamin Straumann, *Roman Law in the State of Nature*, trans. Belinda Cooper (Cambridge: Cambridge University Press, 2015), 154.

p. 135   **"nobody's thing."** *The Digest of Justinian*, vol. 4, trans. and ed. Alan Watson (Philadelphia: University of Pennsylvania Press, 1985), book 41, chap. 1, 1.

p. 136   **ultimately ruled for Pierson.** Pierson v. Post, 3 Cai. R. 175 (1805).

p. 136   **Indigenous peoples lived there.** Mattias Åhrén analyzes how colonizers squared the doctrine of terra nullius with existing international legal principles and political theories to make sense of disposing Indigenous peoples of their land. See Mattias Åhrén, *Indigenous Peoples' Status in the International Legal System* (Oxford: Oxford University Press, 2016), 7–38.

p. 138   **UNCLOS in practice.** Mary Durfee and Rachael Lorna Johnstone, *Arctic Governance in a Changing World* (Lanham, MD: Rowman & Littlefield, 2019), 126–27.

p. 138   **distinct rights and regulations.** The discussion that follows is based on a simplification of the Law of the Sea.

p. 138   **managing the Arctic.** The Ilulissat Declaration, May 28, 2008.

p. 139   **Water Resources Act.** An Act Respecting the Control and Management of Water Resource in the Province, SNL2002, Chapter W-4.01.

p. 141   **environment in Svalbard.** Act of June 15, 2001, No. 29, Relating to the Protection of the Environment in Svalbard.

p. 141   **Australian Environmental Protection and Biodiversity**
          Conservation Act. Environment Protection and
          Biodiversity Conservation Act 1999, No. 91, 1999. This
          point was made by Alex Gardner, professor in the School
          of Law at the University of Western Australia, cited in
          Jackson Gothe-Snape and Emma Machan, "Why a Middle
          Eastern Business Thirsty for Water Can't Just Tow an
          Iceberg from Antarctica," *ABC News*, August 14, 2019,
          https://www.abc.net.au/news/2019-08-14/why-a-middle
          -eastern-business-cant-just-tow-antarctica-iceberg
          /11318638.

p. 142   **regional fishery management organizations.** UNCLOS,
          Article 118.

p. 142   **prohibiting new claims from being made.** ATS, Articles
          1 and 4.

p. 143   **Madrid Protocol.** Protocol on Environmental Protection to
          the Antarctic Treaty, opened for signature October 4, 1991,
          2941 UNTS 3, entered into force January 14, 1998.

p. 143   **"impacts on the Antarctic environment."** Ibid., Article 2.

p. 143   **unless for scientific research, are prohibited.** Ibid.,
          Article 7.

p. 143   **"glacial or marine environments."** Ibid., Article 3.2(b)(iii).

p. 143   **the unique Antarctic environment.** Recommendation
          XV-21 (ATCM XV–Paris, 1989). Prior to the enactment
          of the Madrid Protocol, parties to the ATS addressed
          the environmental impact of iceberg exploitation in
          a 1989 Antarctica Treaty Consultative Meeting, in
          "Recommendation XV-21." Because the recommendation
          went through the formal review process outlined in
          Article IX(4) of the Antarctic Treaty, the document is
          considered binding. The recommendation expresses the
          parties' "concern that uncontrolled activities relating to
          the exploitation of Antarctic icebergs could . . . have an
          adverse effect on the unique Antarctic environment and
          its dependent and associated ecosystems" and affirms that
          "sufficient scientific information is not yet available on the
          environment impacts" if iceberg harvesting were to occur.
          On that basis, the parties recognize "the desirability that
          commercial exploitation of Antarctic ice not occur" prior to
          an examination of those issues by the contracting parties
          to the Antarctic Treaty. Recommendation, https://www

.ats.aq/devAS/Meetings/Measure/190. For more on the binding nature of the recommendation, see Piotr Szwedo, *Cross-Border Water Trade: Legal and Interdisciplinary Perspectives* (Leiden: Brill Nijhoff, 2019), 288.

p. 144   **common-pool resource problem.** For a standard definition, see Elinor Ostrom, *Governing the Commons: The Evolution of Institutions for Collective Action* (Cambridge: Cambridge University Press, 2015), 30–33. For a good discussion of climate change as a common-pool resource problem, see David Morrow, *Values in Climate Policy* (London: Rowman & Littlefield, 2020), 32–42.

p. 145   **tragedy of the commons might result.** Garret Hardin "The Tragedy of the Commons," *Science* 162, no. 3859 (1968): 1243–48.

p. 146   **about how to treat icebergs.** See, for example, Cory Lewis, "Iceberg Harvesting: Suggesting a Federal Regulatory Regime for a New Freshwater Source," *Boston College Environmental Affairs Law Review* 42 (2015): 439–71, and Bryan S. Geon, "A Right to Ice? The Application of International and National Water Laws to the Acquisition of Iceberg Rights," *Michigan Journal of International Law* 19 (1997): 277–99.

p. 147   **if we were to consider them water.** United Nations Convention on the Law of the Non-Navigational Uses of International Watercourses, G.A. Res. 51/229, at 1, UN Doc. A/RES/51/206, May 21, 1997, entered into force August 17, 2014.

p. 147   **among other factors.** Ibid., Article 5, 6.1(a-g), 20.

p. 147   **and work cooperatively.** Ibid., Article 6.2, Article 8.

p. 147   **to vital human needs.** Ibid., Article 10.2.

p. 147   **life-threatening dehydration.** For a good analysis, see Paul Stanton Kibel, "Damage to Fisheries by Dams: The Interplay between International Water Law and International Fisheries Law," *UCLA Journal of International Law & Foreign Affairs* 121 (2017): 140.

p. 148   **exclusive economic zone.** UNCLOS, Article 56.1(a).

p. 148   **exercise exclusive sovereignty over icebergs.** UNCLOS, Article 137.1.

p. 149   **interests of other states.** UNCLOS, Articles 87, 88.

p. 149   **"special requirements of developing States."** UNCLOS, Articles 118, 119.

p. 149   **the resource under UNCLOS.** UNCLOS, Article 66.1.

p. 149   **or a total allowable catch.** UNCLOS, Article 66.2.

p. 150   **"normal catch."** UNCLOS, Article 66.3(a).

p. 150   **of the respective states.** UNCLOS, Article 69.2(d).

p. 150   **"fished in the zone."** UNCLOS, Article 69.4.

p. 151   **as one US court pronounced.** R.M.S. Titanic, Inc. v. Wrecked & Abandoned Vessel, 435 F.3d 521 (2006).

p. 151   **maximize its economic potential.** As noted by Craig Forrest, *International Law and the Protection of Cultural Heritage* (Abingdon, UK: Routledge, 2010), 288.

p. 151   **humankind as a whole.** UNCLOS, Article 136.

p. 151   **state or country of origin.** Today, the traditional "law of finds" is read together with UNCLOS, even though the latter was not designed to regulate such maritime property. In this case, Article 194 would apply.

p. 152   **to mine the resources.** Peter Bautista Payoyo, *Cries of the Sea: World Inequality, Sustainable Development and the Common Heritage of Humanity* (The Hague: Kluwer Law International, 1997), 259.

p. 153   **has not been uncontroversial.** For more on the creation of the International Seabed Authority, see Aline L. Jaeckel, *The International Seabed Authority and the Precautionary Principle: Balancing Deep Seabed Mineral Mining and Marine Environmental Protection* (Leiden: Brill Nijhoff, 2017), 73–87.

p. 153   **well-being of the environment.** For recent criticisms, see Karen McVeigh, "Seabed Regulator Accused of Deciding Deep Sea's Future 'Behind Closed Doors,'" *Guardian*, April 1, 2022, https://www.theguardian.com/environment/ 2022/apr/01/worlds-seabed-regulator-accused-of-reckless -failings-over-deep-sea-mining.

p. 155   **state practice and *opinio juris*.** Hugh Thirlway has written a helpful primer on the subject. See Hugh Thirlway, *The Sources of International Law* (Oxford: Oxford University Press, 2014), 53–92.

p. 155   **shaped much of space law.** Michael P. Scharf, *Customary International Law in Times of Fundamental Change: Recognizing Grotian Moments* (Cambridge: Cambridge University Press, 2013), 37–41.

p. 155   **"rule of law requiring it."** North Sea Continental Shelf (Ger. v. Den.; Ger. v. Neth.), Judgment, I.C.J. Reports, p. 3 (February 20, 1969). Not all state practices rise to the level of a norm of customary law, since states may take actions

for nonlegal reasons—that is, not because they believe
they have a legal duty to do so. Discerning *opinio juris* is
therefore not an uncomplicated exercise. Through internal
and external official statements, however, states often
clarify which behaviors they deem legally obligatory.

p. 157   **illness linked to poor water and sanitation.** World Health
Organization, "Drinking-Water Key Facts," March 21,
2022, https://www.who.int/news-room/fact-sheets/detail/
drinking-water.

p. 157   **as the global population surges.** As the Organisation for
Economic Co-operation and Development estimates. See
*OECD Environmental Outlook to 2050: The Consequences of
Inaction,* 2012, https://read.oecd-ilibrary.org/environment/
oecd-environmental-outlook-to-2050_9789264122246
-en. Scientists, however, have argued that the demand
will actually be greater than many international
organizations predict. See Alberto Boretti and Lorenzo
Rosa, "Reassessing the Projections of the World Water
Development Report," *npi Clean Water,* 2 no. 15 (2019),
https://www.nature.com/articles/s41545-019-0039-9.

p. 158   **Peru to the United States.** Depressingly, the chronology
is updated biennially. For a complete list, see the Pacific
Institute's interactive website: https://www.worldwater.org/
water-conflict.

p. 158   **"fought over water."** Cited in Fredrik Segerfeldt, *Water for
Sale: How Business and the Market Can Resolve the World's
Water Crisis* (Washington, DC: Cato Institute, 2005), 38.

p. 158   **private property schemes.** Elinor Ostrom, *Governing the
Commons: The Evolution of Institutions for Collective Action*
(Cambridge: Cambridge University Press, 1990).

p. 159   **emerge to govern icebergs.** Scholars of customary norms
expect them to form in small tight-knit communities. For
a classic study, I recommend Robert C. Ellickson, *Order
without Law: How Neighbors Settle Disputes* (Cambridge,
MA: Harvard University Press, 1991).

p. 161   **geoengineering schemes.** Oliver Morton's book helped me
to think about geoengineering differently and the good it
might bring about. See Oliver Morton, *The Planet Remade:
How Geoengineering Could Change the World* (Princeton, NJ:
Princeton University Press, 2017).

p. 164   **Greek city-states.** "All the Greeks who were concerned about the general welfare of Hellas met in conference and exchanged guarantees. They resolved in debate to make an end of all their feuds and wars against each other, whatever the cause from which they arose. . . . This they did in the hope that since the danger threatened all Greeks alike, all of Greek blood might unite and work jointly for one common end." See Herodotus, *Histories*, book 7, chap. 154, sec. 1–2.

p. 165   **international collaboration.** "Water Is 'Catalyst' for Cooperation, Not Conflict, UN Chief Tells Security Council," UN News, June 6, 2017, https://news.un.org/en/story/2017/06/558922-water-catalyst-cooperation-not-conflict-un-chief-tells-security-council.

p. 165   **"nor any drop to drink."** Samuel Taylor Coleridge, *The Major Works*, ed. H. J. Jackson (Oxford: Oxford University Press, 1985), 52.

p. 165   **turned to icebergs.** For a lovely analysis see Siobhan Carroll, *An Empire of Air and Water: Uncolonizable Space in the British Imagination, 1750–1850* (Philadelphia: University of Pennsylvania Press, 2015), 73–42.

p. 166   **"the ice was all around."** Coleridge, *Major Works*, 51.

p. 169   **spread across its fronds.** Elizabeth Becker, *Overbooked: The Exploding Business of Travel and Tourism* (New York: Simon & Schuster, 2013), 178.

p. 169   **water in the air in an instant.** For more on the stupendous fountain, see Ed Scott-Clarke and Tom Page, "Dubai Fountain: The Story behind Its Dancing Waters," CNN, February 15, 2019, https://www.cnn.com/travel/article/dubai-fountain-story/index.html.

p. 170   **promotional video depicts.** The Burj Khalifa Fact sheet includes more mind-boggling records held by the building: https://www.burjkhalifa.ae/img/FACT-SHEET.pdf.

p. 170   **per capita in the world.** Stephen Leahy and Katherine Purvis, "Peak Salt: Is the Desalination Dream over for the Gulf States?" *Guardian*, September 29, 2016. Though environmental concerns have long accompanied desalination efforts in the Gulf, at least one recent study suggests there may be less cause for concern than previously thought. Consider, for example, Mohamed A. Dawoud and Mohamed M. Al Mulla, "Environmental Impacts of Seawater Desalination: Arabian Gul Case Study," *International Journal of Environment and Sustainability* 1, no. 3 (2012): 22–37, and Hamed D. Ibrahim, Pengfei Xue, and Elfatih A. B. Eltahir, "Multiple Salinity Equilibria and Resilience of Persian/Arabian Gulf Basin Salinity to Brine Discharge," *Frontiers in Marine Science*, July 10, 2020.

p. 170   **and other international organizations.** Human Rights Watch, "United Arab Emirates, Events of 2021," https:// www.hrw.org/world-report/2022/country-chapters/ united-arab-emirates.

p. 170   **had their passports confiscated.** Uluç Kadıoğlu, "Taken Hostage in the UAE," *Harvard International Review*, July 29, 2022, https://hir.harvard.edu/taken-hostage-in-the-uae. For more on the protests and the legal system that made the abuses possible, see Yasin Kakande, *Slave States: The Practice of Kafala in the Gulf Arab Region* (Winchester, UK: Zero Books, 2015).

p. 171   **Beatrice leads Dante through heaven.** "My guide and I came on that hidden road / to make our way back into the bright world; / and with no care for any rest, we climbed— / he first, I following—until I saw, / through a round opening, some of those things / of beauty Heaven bears. It was from there/ that we emerged, to see—once more—the stars." Dante Alighieri, *Inferno*, 317.

p. 171   **industry is continuing to grow.** Manzoor Qadir and Vladimir Smakhtin, "Where the Water Is," *Project Syndicate*, May 17, 2018, https://www.project-syndicate.org/ commentary/tapping-unconventional-freshwater-sources -by-manzoor-qadir-and-vladimir-smakhtin-2018-05.

p. 171   **most need freshwater.** Alister Doyle, "Too Much Salt: Water Desalination Plants Harm Environment: U.N.," Reuters, January 14, 2019, https://www.reuters .com/article/us-environment-brine/too-much-salt -water-desalination-plants-harm-environment-u-n -idUSKCN1P81PX; Joe McCarthy and Erica Sánchez, "This Clean Water Solution Actually Produces Trillions of Gallons of Toxic Sludge Worldwide," Global Citizen, January 15, 2019, https://www.globalcitizen.org/en /content/desalination-water-toxic-sludge.

p. 172   **to use at home.** Daniel A. Gross, "Recycling Sewage into Drinking Water Is No Big Deal. They've Been Doing It in Namibia for 50 Years," *The World*, December 15, 2016, https://theworld.org/stories/2016-12-15/recycling-sewage

-drinking-water-no-big-deal-theyve-been-doing-it-namibia
-50-years.

p. 172   **embracing the final product.** See: David M. Glick, Jillian
            L. Goldfarb, Wendy Heiger-Bernays, and Douglas L.
            Kriner, "Public Knowledge, Contaminant Concerns, and
            Support for Recycled Water in the United States," *Resources,
            Conservation and Recycling* 150 (2019), https://www
            .sciencedirect.com/science/article/pii/S0921344919303143;
            Julia Wester, Kiara R. Timpano, Demet Çek, and
            Kenneth Broad, "The Psychology of Recycled Water:
            Factors Predicting Disgust and Willingness to Use," *Water
            Resources Research* 52, no. 4 (2016): 3212–26.

p. 172   **daily water needs.** Tom Gould, "Selling Water at $150/m3
            to the World's Poorest People—with Billionaire Backing,"
            Atmospheric Water Generation 21, no. 5 (May 21, 2020),
            https://www.globalwaterintel.com/global-water-intelligence
            -magazine/21/5/general/selling-water-at-150-m3-to-the
            -world-s-poorest-people-with-billionaire-backing.

p. 173   **"accept the first story they hear."** Thucydides, *History*, 47.

Index